Humidity Sensors

Humidity Sensors

Advances in Reliability, Calibration and Application

Special Issue Editors

Peter W. McCarthy
Zhuofu Liu
Vincenzo Cascioli

MDPI • Basel • Beijing • Wuhan • Barcelona • Belgrade

MDPI

Special Issue Editors

Peter W. McCarthy
University of South Wales
UK

Zhuofu Liu
Harbin Univesity of Science and Technology
China

Vincenzo Cascioli
Murdoch University
Australia

Editorial Office
MDPI
St. Alban-Anlage 66 4052
Basel, Switzerland

This is a reprint of articles from the Special Issue published online in the open access journal *Sensors* (ISSN 1424-8220) from 2018 to 2019 (available at: https://www.mdpi.com/journal/sensors/special_issues/humidity_sensors)

For citation purposes, cite each article independently as indicated on the article page online and as indicated below:

LastName, A.A.; LastName, B.B.; LastName, C.C. Article Title. *Journal Name* **Year**, *Article Number*, Page Range.

ISBN 978-3-03921-122-7 (Pbk)
ISBN 978-3-03921-123-4 (PDF)

Contents

About the Special Issue Editors

Peter W. McCarthy obtained a BSc jt. Hons in Physiology and a PhD in Neurophysiology from the University of Manchester and the University of St Andrews, respectively. He has valuable experience assessing the activity of the body and its component systems. His awareness for measurement accuracy issues in clinical technology was first raised while working on ear thermometry with the UK's National Physical Laboratory. He was awarded a full professorship of Clinical Technology at the University of Glamorgan in 2008. His current interests surround the use of technology to better understand the role of neurophysiological sensory feedback mechanisms, with the aim to eventually create intelligent replacements for those with sensory deficits. This includes relating perceptions of the person to body-seat interface parameters, assessing and preventing cervical spine dysfunction in elite sports and optimizing brain-computer interfacing.

Zhuofu Liu received his Masters and PhD from Harbin Engineering University, Harbin, China, in 2001 and 2004, respectively. In 2005 he served as an associate professor at the School of Underwater Acoustic Engineering, Harbin Engineering University. In 2006 he worked as an academic visitor at the University of Oxford. From 2007 to 2009 he worked as a research associate at the Welsh Institute of Chiropractic, University of Glamorgan (now University of South Wales), Pontypridd, UK. Since 2010 he has been a professor at the School of Measurement Control and Communication Engineering, Harbin University of Science and Technology. His research interests include image processing, biomedical signal acquisition and analysis, and healthcare information technology. Dr. Liu is currently the principal investigator for several projects investigating the body-seat interface microenvironment.

Vincenzo Cascioli obtained a Masters in Chiropractic from Durban University of Technology, South Africa and a PhD in Ergonomics from the University of South Wales, UK. His current research interests involve the use of technology to evaluate the factors, such as temperature, humidity and movement, associated with sitting comfort or discomfort.

Preface to "Humidity Sensors"

This Special Issue, "Humidity Sensors: Advances in Reliability, Calibration and Application", contains a range of articles illustrating the growth in use and form of humidity sensors. It is obvious from the contents of this volume that humidity detection has come a long way since wet bulb psychrometry. The number of electronic sensor-based methods available for detecting and reporting relative humidity appears to have grown exponentially. However, as one moves further away from the physical measurement of a property, issues of reliability and accuracy of calibration become increasingly important. In the case of humidity, the property of a sensor that enables measurements to be made can also be the property that leads to issues with calibration and sensitivity, as well as recovery of the sensor. All of these factors may limit the uptake and application of the sensors. This volume is a window into the recent, rapid growth in research aimed at finding the best method for sensing humidity in fields ranging from biomedicine, agriculture, and pharmacology to semiconductors and food processing. Never has there been a greater need to study and refine these sensors.

In our contribution the editors have taken the opportunity to follow up on colleagues' questions regarding the source of spurious and short lived, but potentially vital, artifacts associated with one potential use of humidity sensors: assessing seating or mattress breathability. For this, we have gone back to basics to illustrate the effects a delay in the equilibration of temperature at the sensor site can have on the sensor's reporting of relative humidity in the surrounding environment. This relatively minor artifact shows how believing without questioning can mislead and obfuscate, whereas questioning can open new areas for development.

We initially considered this a good point in time to bring together available research (potential and actual) and look at the issues surrounding this measurement. This issue shows the breadth of use and hints at the future potential of these sensors.

Peter W. McCarthy, Zhuofu Liu, Vincenzo Cascioli
Special Issue Editors

sensors

MDPI

Article

Determination of Optimal Measurement Points for Calibration Equations—Examples by RH Sensors

Hsuan-Yu Chen [1] and Chiachung Chen [2,*]

1   Department of Materials Science and Engineering, University of California, San Diego, CA 92093, USA;
    wakaharu37@gmail.com
2   Department of Bio-Industrial Mechatronics Engineering, National ChungHsing University,
    Taichung 40227, Taiwan
*   Correspondence: ccchen@dragon.nchu.edu.tw; Tel.: +886-4-2285-7562

Received: 26 February 2019; Accepted: 6 March 2019; Published: 9 March 2019

Abstract: The calibration points for sensors must be selected carefully. This study uses accuracy and precision as the criteria to evaluate the required numbers of calibration points required. Two types of electric relative humidity (RH) sensors were used to illustrate the method and the standard RH environments were maintained using different saturated salt solutions. The best calibration equation is determined according to the *t*-value for the highest-order parameter and using the residual plots. Then, the estimated standard errors for the regression equation are used to determine the accuracy of the sensors. The combined uncertainties from the calibration equations for different calibration points for the different saturated salt solutions were then used to evaluate the precision of the sensors. The accuracy of the calibration equations is 0.8% RH for a resistive humidity sensor using 7 calibration points and 0.7% RH for a capacitance humidity sensor using 5 calibration points. The precision is less than 1.0% RH for a resistive sensor and less than 0.9% RH for a capacitive sensor. The method that this study proposed for the selection of calibration points can be applied to other sensors.

Keywords: calibration points; saturated salt solutions; humidity sensors; measurement uncertainty

1. Introduction

The performance of sensors is key for modern industries. Accuracy and precision are the most important characteristics. Calibration ensures sensors' performance. When a sensor is calibrated, the reference materials or reference environments must be specified. For a balance calibration, a standard scale is the reference materials. For temperature calibration, the triple point of ice-water or boiling matter is used to maintain the reference environment.

The experimental design for calibration must consider the following factors [1–3].

1.   The number and the location of the calibration points.
2.   The regression equations (linear, poly-nominal, non-linear).
3.   The regression techniques.
4.   The standard references and their uncertainties.

Betta [1] adopted minimizing the standard deviations for the regression curve coefficients or the standard deviation for the entire calibration curve to design an experiment to determine the number of calibration points, the number of repetitions, and the location of calibration points. Three types of sensor were used to demo the linear, quadratic and cubic calibration equations: a pressure transmitter, a platinum thermometer and E-Type thermocouple wires. The estimated confidence interval values were used to determine the validity of the regression equation. This method was extended to address calibration for complex measurement chains [2].

Hajiyev [3] noted the importance of the selection of the calibration points to ensure the accuracy of the calibration and the optimal selection of standard pressure setters and used an example to verify the method. A dispersion matrix, $\underset{D}{\rightarrow}$ of the estimated coefficients was defined and this matrix $\underset{D}{\rightarrow}$ was used as a scale of the error between the sensor and the reference instruments. Two criteria were used to evaluate the performance. The minimized sum of the diagonal elements of the matrix $\underset{D}{\rightarrow}$ is called the A-optimality criterion. The minimized of the generalized of determinant of the matrix $\underset{D}{\rightarrow}$ is called the D-optimality criterion. The optimal measurement points for the calibration of the differential pressure gages were determined using the A-optimality criterion [3] and the D-optimality criterion [4]. Khan et al. [5] used an inverse modeling technique with a critical neural network (ANN) to evaluate the order of the models and the calibration points. The root-mean-square error (RMSE) was used as the criterion.

Recently, modern regression has been used as an important role to express the quantitative relationship between independent and response variables for tests on a single regression coefficient [6–9]. This technique used to address calibration equations and the standard deviations of these calibration equations then served as the criteria to determine their accuracy [10,11].

The confidence band for the entire calibration curve or for each experimental point was used to evaluate the fit of calibration equations [1,2]. The concept of measurement uncertainty (MU) is widely used to represent the precision of calibration equations [12–14]. Statistical techniques can be used to evaluate the accuracy and precision of calibration equations that are obtained using different calibration points [15–17]. Humidity sensors that were calibrated using different saturated salt solutions were tested to illustrate the technique for the specification of optimal measurement points [18,19].

Humidity is very important for various industries. Many manufacturing and testing processes, such as those for food, chemicals, fuels and other products, require information about humidity [20]. Relative humidity (RH) is commonly used to express the humidity of moist air [21]. Electric hygrometers are the most commonly used sensors because they allow real-time measurement and are easily operated.

The key performance factors for an electrical RH meter are the accuracy, the precision, hysteresis and long-term stability. At high air humidity measurement, there is a problem with response time of the RH sensors in conventional methods. The solution for this problem for high air humidity measurement is to use an open capacitor with very low response time [22–24] and quartz crystals which compensate temperature drift. An environment with a standard humidity is required for calibration. Fixed-point humidity systems that use a number of points with a fixed relative humidity are used as a standard. A humidity environment is maintained using different saturated salt solutions. The points with a fixed relative humidity are certified using various saturated salt solutions [19]. When the air temperature, water temperature and air humidity reach an equilibrium state, constant humidity is maintained in the air space [19].

The RH value that is maintained by the salt solutions is of interest. Wexler and Hasegawa measured the relative humidity that is created by eight saturated salt solutions using the dew point method [25]. Greenspan [18] compiled RH data for 28 saturated salt solutions. The relationship between relative humidity and ambient temperature was expressed as a 3rd or 4th polynomial equation. Young [26] collected RH data for saturated salt solutions between 0 to 80 °C and plotted the relationship between relative humidity and temperature. The Organisation Internationale De Metrologies Legale (OIML) [19] determined the effect of temperature on the relative humidity of 11 saturated salt solutions and tabulated the result. Standard conditions, devices and the procedure for using the saturated salt solutions were detailed.

The range for the humidity measurement is from about 11% to 98% RH. Studies show that the number of fixed-point humidity references that are required for calibration is inconsistent. Lake et al. [27] used five salt solutions for calibration and found that the residuals for the linear calibration equation were distributed in a fixed pattern. Wadso [28] used four salt solutions to determine the RH that was generated in sorption balances. Duvernoy et al. [29] introduced seven salt solutions to generate the RH for a metrology laboratory. Bellhadj and Rouchou [30] recommended five salt solutions and two sulfuric acids to create the RH environment to calibrate a hygrometer.

There is inconsistency in the salt solutions that are specified by instrumentation companies and standard bodies. The Japanese Mechanical Society (JMS) specifies 9 salt solutions for the standard humidity environment [31]. The Japanese Industrial Standards Committee (JISC) recommends 4 salt solutions to maintain RH environment [32]. The Centre for Microcomputer Applications (CMA) company specifies 11 salt solutions [33]. Delta OHM use only 3 salt solutions [34]. The OMEGA company use 9 salt solutions [35]. TA instruments specifies 9 salt solutions [36] and Vaisala B.V. select 4 salt solutions [37]. These salt solutions are listed in Table 1.

Table 1. The selection of saturated salt solutions that are used to calibrate humidity sensors.

Salt Solutions	OIMI [19]	Lake [27]	Wadso [28]	Duvernoy [29]	Belhadj [30]	JMS [31]	JISC [32]	CMA [33]	Delta [34]	OMEGA [35]	TA [36]	Vaisala [37]
LiBr								*				
LiCl	*		*	*	*			*	*	*	*	*
CH_3COOK	*							*		*	*	
$MgCl_2 \cdot GH_2O$	*		*	*	*	*		*	*	*	*	*
K_2CO_3	*			*	*	*		*		*	*	
$Mg(NO_3)_2$		*	*		*	*		*		*	*	
NaBr	*			*	*						*	
KI	*	*						*				
$SrCl_2$											*	
NaCl	*	*	*	*	*	*	*	*	*	*	*	*
$(NH_4)_2SO_4$								*				
KCl	*	*		*	*	*	*	*		*	*	
KNO_3					*	*	*			*		
K_2SO_4	*	*		*	*	*		*		*		*

Note: OIML, The Organisation Internationale De Metrologies Legale.

Lu and Chen [17] calculated the uncertainty for humidity sensors that were calibrated using 10 saturated salt solutions for two types of humidity sensors. The study showed that a second-order polynomial calibration equation gave better performance than a linear equation. The measurement uncertainty is used as the criterion to determine the precision performance of sensors [38].

The number of standard relative humidity values for fixed-point humidity systems is limited by the number and type of salt solutions. The number of salt solutions that must be used to specify the calibration points for the calibration of RH sensors is a moot point. More salt solutions allow more calibration points for the calibration of RH sensors. However, using more salt solutions is time-consuming. This study determined the effect of the number and type of salt solutions on the calibration equations for two types of humidity sensors. The accuracy and precision were determined in order to verify the method for the choice of the optimal calibration points for sensor calibration.

2. Materials and Methods

2.1. Relative Humidity (RH) and Temperature Sensors

Resistive sensor (Shinyei THT-B141 sensor, Shinyei Kaisha Technology, Kobe, Japan) and capacitive sensor (Vaisala HMP-143A sensor, Vaisala Oyj, Helsinki, Finland) were used in this study. The specification of the sensors is listed in Table 2.

Table 2. The specifications of two humidity sensors.

	Resistive Sensor	Capacitive Sensor
Model 1	THT-B121	HMP 140A
Sensing element	Macro-molecule HPR-MQ	HUMICAP
Operating range	0–60 °C	0–50 °C
Measuring range	10–99% RH	0–100%
Nonlinear and repeatability	±0.25% RH	±0.2% RH
ResolutionTemperature effect	0.1% RH (relative humidity)none	0.1% RH0.005%/°C

2.2. Saturated Salt Solutions

Eleven saturated salt solutions were used to maintain the relative humidity environment. These salt solutions are listed in Table 3.

Table 3. The Calibration points for saturated salt solutions to establish the calibration equations.

Salt Solutions	(n_1 = 11) Case 1	(n_2 = 9) Case 2	(n_3 = 7) Case 3	(n_4 = 5) Case 4	u_c
LiCl	*	*	*	*	0.27
CH$_3$COOK	*				0.32
MgCl$_2$	*	*	*	*	0.16
K$_2$CO$_3$	*	*	*		0.39
Mg(NO$_3$)$_2$	*	*			0.22
NaBr	*	*	*	*	0.40
KI	*	*			0.24
NaCl	*	*	*	*	0.12
KCl	*	*	*		0.26
KNO$_3$	*				0.55
K$_2$SO$_4$	*	*	*	*	0.45

Note: u_c values were obtained from Greenspan [18] and The Organisation Internationale De Metrologies Legale (OIML) R121 [19].

2.3. Calibration of Sensors

The humidity probes for the resistive and capacitive sensors were calibrated using saturated salt solutions. A hydrostatic solution was produced in accordance with OIML R121 [19]. The salt was dissolved in pure water in a ratio such that 40–75% of the weighted sample remained in the solid state. These salt solutions were stored in containers.

The containers were placed in a temperature controller at an air temperature of 25 ± 0.2 °C. During the calibration process, humidity and temperature probes were placed within the container above the salt solutions. The preliminary study showed that an equilibrium state is established in 12 h so the calibration lasted 12 h to ensure that the humidity of the internal air had reached an equilibrium state. Experiments for each RH environment were repeated three times. The temperature was recorded and the standard humidity of the salt solutions was calculated using Greenspan's equation [18].

2.4. Establish and Validate the Calibration Equation

The experimental design and flow chart for the data analysis is shown in Figure 1.

The relationship between the standard humidity and the sensor reading values was established as the calibration equation.

This study used the inverse method. The standard humidity is the dependent (y_i) and the sensor reading values are the independent variables (x_i) [17].

The form of the linear regression equation is:

$$Y = b_0 + b_1 X \tag{1}$$

4

where b_0 and b_1 are constants.

The form of the higher-order polynomial equation is:

$$Y = c_0 + c_1 X + c_2 X^2 + c_3 X^3 + \ldots + c_k X^k \tag{2}$$

where c_0, c_1 to c_k are constants.

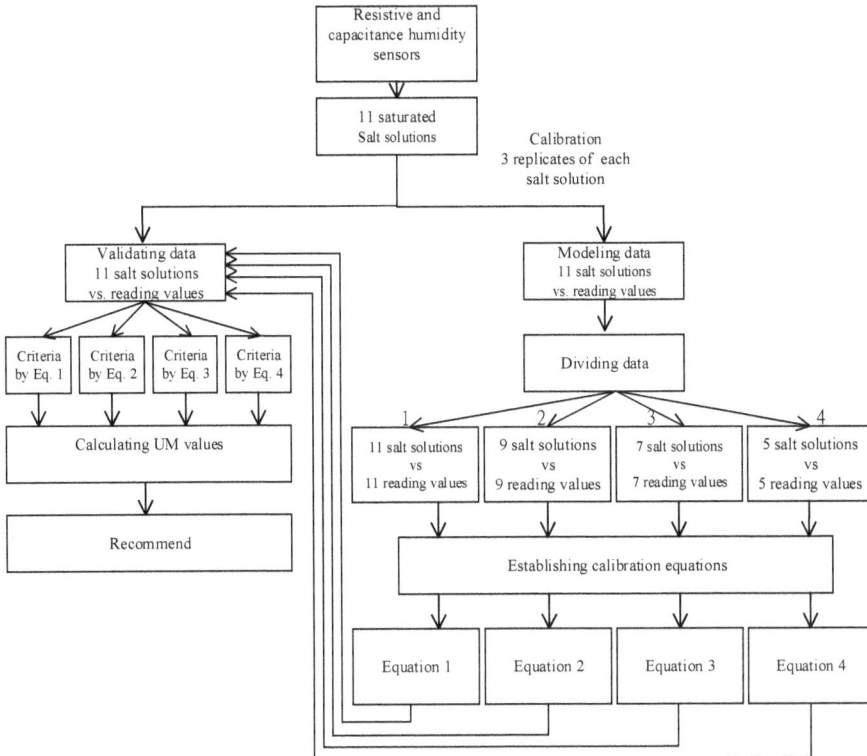

Figure 1. The experimental design and flowchart of data analysis.

2.5. Different Calibration Points

To model the calibration equations, the data for four different salt solutions was used, as listed in Table 3.

Case 1: The data set is for 11 salt solutions and 11 calibration points
Case 2: The data set is for 9 salt solutions and 9 calibration points
Case 3: The data set is for 7 salt solutions and 7 calibration points
Case 4: The data set is for 5 salt solutions and 5 calibration points

For each sensor, four calibration equations were derived using four different calibration points.

2.6. Data Analysis

The software, Sigma plot ver.12.2, was used to determine the parameters for the different orders of polynomial equations.

5

2.6.1. Tests on a Single Regression Coefficient

The criteria to assess the fit of the calibration equations are the coefficient of determination R^2, the estimated standard error of regression s and the residual plots.

The coefficient of determination, R^2 is used to evaluate the fit of a calibration equation. However, no standard criterion has been specified [15,16].

The single parameter coefficient was tested using the t-test to evaluate the order of polynomial regression equation. The hypotheses are:

$$H_0 : b_k = 0 \tag{3}$$

$$H_1 : b_k \neq 0 \tag{4}$$

The t-value is:

$$t = b_k / se(b_k) \tag{5}$$

where b_k is the value of the parameter for the polynomial regression equation of the highest order, and $se(b_k)$ is the standard error of b_k.

2.6.2. The Estimated Standard Error of Regression

The estimated standard error of regression s is calculated as follows:

$$s = \left(\frac{(\hat{y}_2 - y_i)^2}{n_1 - p} \right)^{0.5} \tag{6}$$

where $\hat{y}_i$ is the predicted valued of the response, $\hat{y}_i$ is the response, n_1 is the number of data and p is the number of parameters.

The s value is the criterion that is used to determine the accuracy of a calibration equations [38]. It is used to assess the accuracy of two types of RH sensors that are calibrated using different saturated salt solutions.

2.6.3. Residual Plots

Residual plots is the quantitative criterion that is used to evaluate the fit of a regression equation. If the regression model is adequate, the data distribution for the residual plot should tend to a horizontal band and is centered at zero. If the regression equation is not accepted, the residual plots exhibit a clear pattern.

For the calibration equation, tests on a single regression coefficient and the residual plots are used to determine the suitability of a calibration equation for RH sensors that are calibrated using different saturated salt solutions. The estimated standard error of the regression equations is then used to determine the accuracy of the calibration equations.

2.7. Measurement Uncertainty for Humidity Sensors

The measurement uncertainty for RH sensors using different salt solutions was calculated using International Organization for Standardization, Guide to the Expression of Uncertainty in Measurement (ISO, GUM) [12,13,17].

$$u_c^2 = u^2 x_{pred} + u^2_{temp} + u^2_{non} + u^2_{res} + u^2_{sta} \tag{7}$$

where u_c is the combined standard uncertainty, ux_{pred} is the uncertainty for the calibration equation, u_{temp} is the uncertainty due to temperature variation, u_{non} is the uncertainty due to nonlinearity, u_{res} is the uncertainty due to resolution, and u_{sta} is the uncertainty of the reference standard for the saturated salt solution.

The uncertainty of x_{pred} is calculated as follows [38]:

$$ux_{pred} = s\sqrt{1 + \frac{1}{n} + \frac{(y - \bar{y})^2}{\sum(yi^2) - \frac{(\sum y_i)^2}{n}}} \qquad (8)$$

where $\bar{y}$ is the average value of the response.

The uncertainty in the value of u_{ref} for the saturated salt solutions is determined using the reference standard for the salt solution. The scale and the uncertainty of these saturated salt solutions are listed in Table 3 that are taken from Greenspan [18] and the Organisation Internationale De Metrologies Legale (OIML) R121 [19]:

$$u_{ref} = \left(\frac{\sum(u_{ri})^2}{N_2}\right)^{0.5} \qquad (9)$$

where u_{ri} is the uncertainty in the humidity for each saturated salt solution and N_2 is the number of saturated salt solutions that are used for calibration.

The calibration equations use different numbers of saturated salt solutions had its uncertainty. This criterion is used to evaluate the precision of RH sensors.

The accuracy and precision of RH sensors that are calibrated using different saturated salt solutions was determined using the s and u_c values. By Equations (7)–(9), the contrast between the number of saturated salt solutions is considered. The greater the number of data points that are used, the smaller is the s value that is calculated by Equation (6). However, this requires more experimental time and cost and the value of u_{ref} may be increased. The uncertainty of each calibration point is different because different saturated salt solutions are used. The optimal number of calibration points were evaluated by accuracy and precision.

3. Results and Discussion

3.1. The Effect of the Accuracy of Different Calibration Points

3.1.1. THT-B121 Resistive Humidity Sensor

Calibration equations for resistive sensors using 11 salt solutions:

The distribution of the relative humidity data for the reading values for a resistive sensor is plotted against the standard humidity values that are maintained using 11 saturated salt solutions in Figure 2.

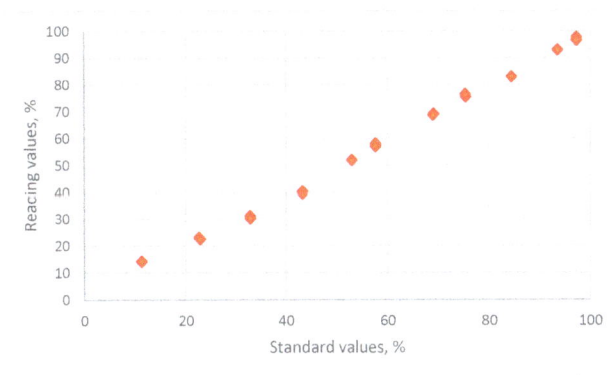

Figure 2. The distribution of the relative humidity data for reading values versus the standard humidity values for THT-B121 resistive humidity sensor using 11 saturated salt solutions (LiCl, CH$_3$COOK, MgCl$_2$, K$_2$CO$_3$, Mg(NO$_3$)$_2$, NaBr, KI, NaCl, KCl, KNO$_3$ and K$_2$SO$_4$).

The estimated parameters and the evaluation criteria for regression analysis are listed in Table 4. The residual plots for the calibration equations for different orders of polynomial equations are shown in Figure 3.

Table 4. Estimated parameters and evaluation criteria for the linear and several polynomial equations for THT-B121 resistive sensor using 11 salt solutions.

	Linear	2nd Order	3nd Order	4th Order
b_0	0.028672	−2.74999	−11.0702	−20.5303
b_1	1.008985	1.13766	1.780025	2.805196
b_2		−0.0011437	−0.01432	−0.0491534
b_3			7.81681×10^{-5}	5.39281×10^{-4}
b_4				-2.07539×10^{-6}
R^2	0.9967	0.9974	0.9987	0.9993
s	1.6098	1.4612	0.982	0.7719
Residual plots	clear pattern	clear pattern	clear pattern	uniform distribution

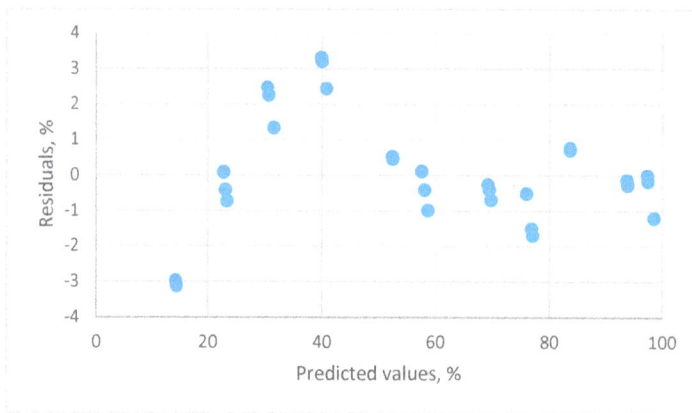

(**a**) Linear equation

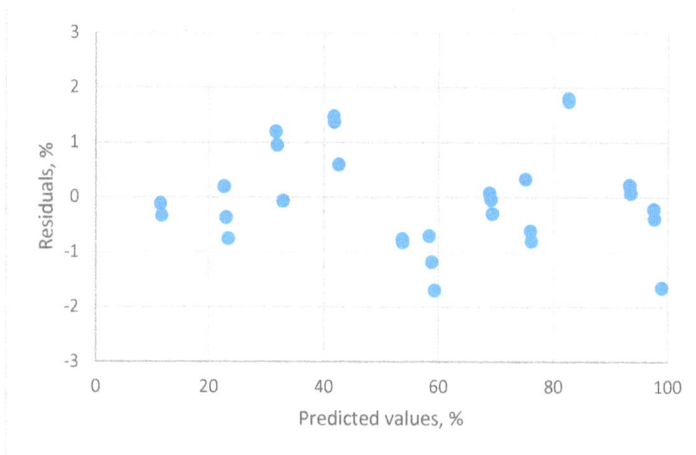

(**b**) 2nd polynomial equation

Figure 3. *Cont.*

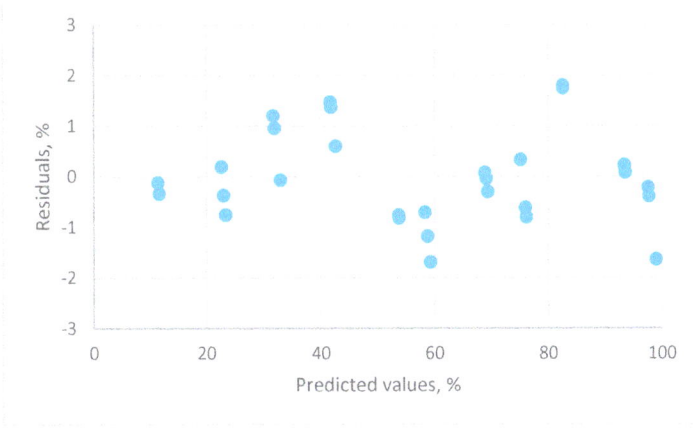

(c) 3rd polynomial equation

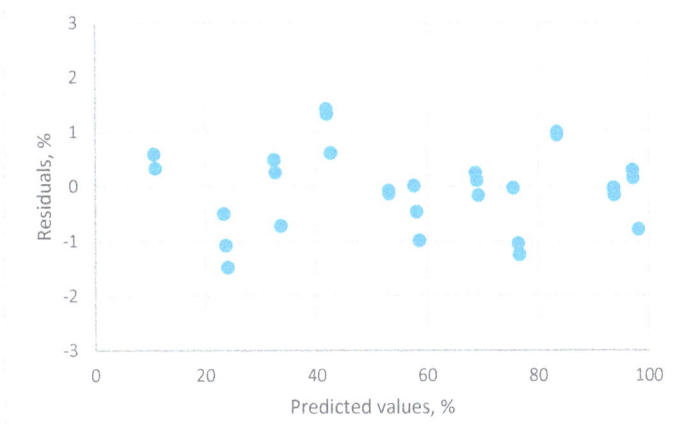

(d) 4th polynomial equation

Figure 3. The residual plots for the calibration equations for different orders of polynomial equations for THT-B121 resistive humidity sensor using 11 saturated salt solutions (LiCl, CH₃COOK, MgCl₂, K₂CO₃, Mg(NO₃)₂, NaBr, KI, NaCl, KCl, KNO₃ and K₂SO₄).

The linear (Figure 3a), 2nd (Figure 3b) and 3rd (Figure 3c) order polynomial equations all exhibit a systematic distribution of residuals. These equations were not satisfactory for resistive sensors. The distribution of residual plots for the 4th order polynomial equations exhibit a uniform distribution (Figure 3d). The t-value for the highest-order parameter ($b_4 = -2.07539 \times 10^{-6}$) was significantly different to zero, so the 4th order polynomial equation is the only adequate calibration equation. The equation is:

$$y = -20.530298 + 2.805196x - 0.049153x^2 + 0.000539x^3 - 2.07539 \times 10^{-6}x^4$$
$$(s_b = 2.5004 \; s_b = 0.2590 \; s_b = 0.0082 \; s_b = 0.00016 \; s_b = 4.770 \times 10^{-7}$$
$$t = -8.2107 \; t = 11.181 \; t = -6.005 \; t = -5.0663 \; t = -4.3514)$$
$$R^2 = 0.992, \; s = 0.7719$$

The coefficient of determination, R^2, for the linear, 2nd, 3rd and 4th order polynomial calibration equations are 0.9967, 0.9974, 0.9987 0.9993, respectively. High R^2 values do not give useful information for the specification of an appropriate calibration equation. The estimated values of standard deviation, s, is used to define the uncertainty for an inverse calibration equation [35]. The s values for the four calibration equations are 1.6098, 1.4612, 0.9820 and 0.7719, respectively. It is seen that an appropriate calibration equation gives a significant reduction in uncertainty.

Calibration equations for resistive sensor using 5 salt solutions:

The estimated parameters and the evaluation criteria for the regression analysis for 5 calibration points for a resistive sensor are listed in Table 5. The residual plots for four calibration equations are shown in Supplementary Materials. Similarly to the regression results for 11 salt solutions, the linear, 2nd and 3rd order polynomial equations all employed a systematic distribution in the residuals plots. These equations are clearly not appropriate calibration equations. For a resistive sensor, the residual plots for the 4th order polynomial equations presented a random distribution.

Table 5. Estimated parameters and evaluation criteria for the linear and several polynomial equations for THT-B121 resistive sensors using 5 salt solutions.

	Linear	2nd Order	3nd Order	4th Order
b_0	−0.970118	−3.1191770	−12.201481	−19.471802
b_1	1.0155235	1.12632754	1.8869907	2.743833
b_2		−0.001007316	−0.01685101	−0.04766345
b_3			9.34623×10^{-5}	5.15689×10^{-4}
b_4				$−1.93676 \times 10^{-6}$
R^2	0.9969	0.9974	0.9994	0.9991
s	1.8109	1.7146	0.7984	1.084
Residual plots	clear pattern	clear pattern	clear pattern	uniform distribution

The R^2 values for the linear, 2nd, 3rd and 4th order polynomial calibration equations are 0.9969, 0.9974, 0.9994 and 0.9998, respectively. However, these higher R^2 values do not provide relevant information about the calibration equations. The s values represent the uncertainty of calibration equations. For the linear, 2nd, 3rd and 4th order polynomial calibration equations are 1.8109, 1.7146, 0.7954 and 1.084, respectively. The 4th order polynomial equations is:

$$y = -19.471802 + 2.743833x - 0.047663x^2 + 0.0005157x^3 - 1.93676 \times 10^{-6}x^4$$
$$(s_b = 2.2789 \; s_b = 0.25086 \; s_b = 0.00869 \; s_b = 0.000117 \; s_b = 5.360 \times 10^{-7}$$
$$t = -8.5447 \; t = 10.9396 \; t = -5.4849 \; t = 4.3946 \; t = -3.6101)$$
$$R^2 = 0.991, s = 1.014$$

The regression results for the 4th order polynomial equations using different calibration points in different salt solutions are listed in Table 6. The results for 9 and 7 calibration points are similar to those for 11 and 5 calibration points.

Table 6. Estimated parameters and evaluation criteria for the 4th order polynomial equations for THT-B121 resistive sensors using four different calibration points.

	Case 1 ($n_1 = 11$)	Case 2 ($n_2 = 9$)	Case 3 ($n_3 = 7$)	Case 4 ($n_4 = 5$)
b_0	−20.530297	−23.41845561	−23.904948	−19.4718019
b_1	2.8051965	3.5861653	3.243023015	2.743832845
b_2	−0.04915334	−0.06230766	−0.06426625	−0.047663446
b_3	5.39281×10^{-4}	7.0951×10^{-4}	7.34202×10^{-4}	5.15689×10^{-4}
b_4	$−2.07539 \times 10^{-6}$	$−2.81734 \times 10^{-6}$	$−2.92042 \times 10^{-6}$	$−1.93676 \times 10^{-6}$
R^2	0.9993	0.9994	0.9994	0.9991
s	0.7719	0.6951	0.8039	1.084

The R^2 value is used b to evaluate the calibration equations [27,33]. Even the linear calibration equation for this study shows a high R^2 value. However, the estimated error was higher than that for other equations. The residual plots all exhibited a clear pattern distribution so the R^2 value cannot be used as the sole criterion to assess the calibration equation. Betta and Dell'Isola [1] mention R^2, Chi-square and F-test to verify the accuracy of a model. This study used t-value for a parameter was used as the criterion. This method bases on statistical theory.

3.1.2. HMP 140A Capacitive Humidity Sensor

Calibration equations for a capacitive sensors using 11 salt solutions

The relationship between the reading values for a capacitive sensor and the standard humidity values that are maintained using 11 saturated salt solutions is shown in Figure 4.

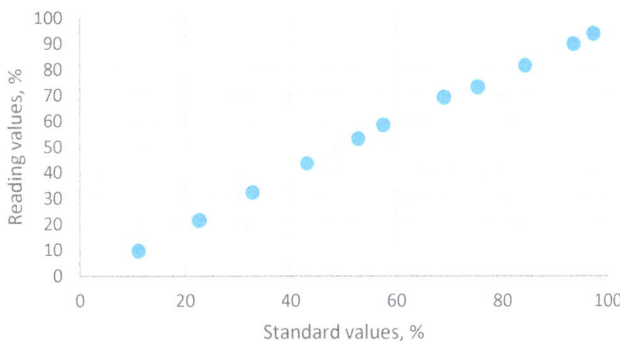

Figure 4. The distributions of relative humidity data for standard humidity values versus the reading values for HMP 140A capacitance humidity sensors using 11 saturated salt solutions (LiCl, CH_3COOK, $MgCl_2$, K_2CO_3, $Mg(NO_3)_2$, NaBr, KI, NaCl, KCl, KNO_3 and K_2SO_4).

The estimated parameters and the evaluation criteria for regression analysis are listed in Table 7.

Table 7. Estimated parameters and evaluation criteria for the linear and polynomial equations for HMP 140A capacitive sensor using 11 salt solutions.

	Linear	2nd Order
b_0	−0.414520	3.479518
b_1	1.031003	0.833274
b_2		0.00186718
R^2	0.9975	0.9994
s	1.4002	0.6837
Residual plots	clear pattern	Uniform distribution

The residual plots for the calibration equations for different orders of polynomial equations are shown in Figure 5.

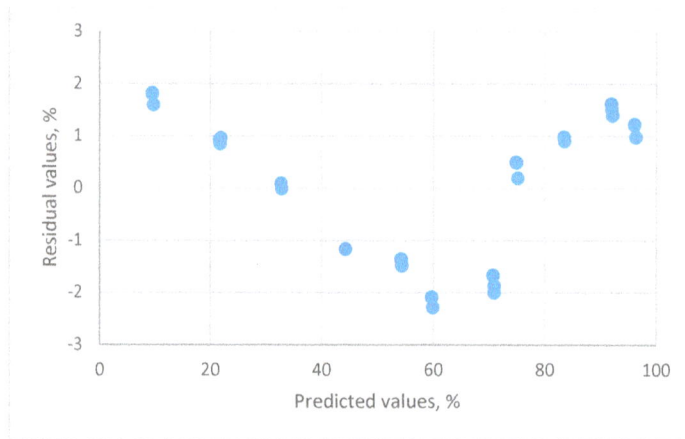

(a) linear equation

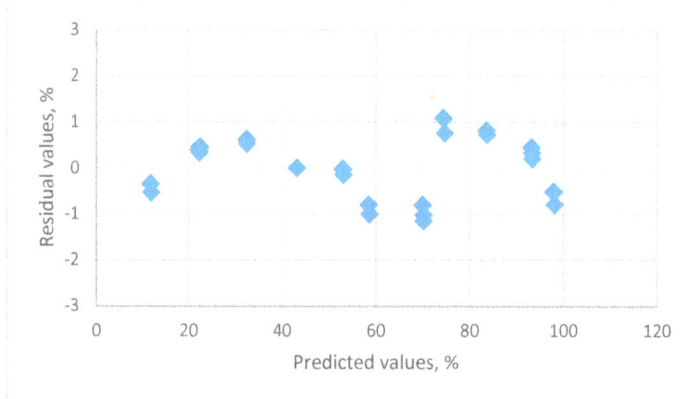

(b) 2ⁿᵈ polynomial equation

Figure 5. The residual plots for the calibration equations for different orders of polynomial equations for HMP 140A capacitance humidity sensor using 11 saturated salt solutions (LiCl, CH_3COOK, $MgCl_2$, K_2CO_3, $Mg(NO_3)_2$, NaBr, KI, NaCl, KCl, KNO_3 and K_2SO_4).

The linear equation (Figure 5a) exhibited a systematic distribution of residuals. The 2nd (Figure 5b) and 3rd (not presented) order polynomial equations both displayed a uniform distribution. The *t*-value for the 3rd order parameter was not significantly different to zero, so the 2nd order polynomial equation is the appropriate calibration equation and list as follows:

$y = 3.479518 + 0.833274x + 0.001867x^2$, $R^2 = 0.9994$, $s = 0.6837$

($s_b = 0.4805$ $s_b = 0.02028$ $s_b = 0.000187$

$t = 7.2408$ $t = 41.098$ $t = 10.004$)

The coefficient of determination, R^2, for the linear and 2nd order polynomial calibration equations are 0.9975 and 0.9994, respectively. The s values for the two calibration equations are 1.4002 and 0.6837, respectively. An appropriate calibration equation gives a significant reduction in the estimated error.

Calibration equations for a capacitive sensor using 5 salt solutions

The estimated parameters and the evaluation criteria for the regression analysis for 5 calibration points for a capacitance are listed in Table 8. The residual plots for four calibration equations are shown in Supplementary Materials. Similarly to the regression results for 11 salt solutions, residuals plots for the linear equation exhibit a systematic distribution. Residual plots for the 2nd order polynomial equations presented a random distribution.

Table 8. Estimated parameters and evaluation criteria for the linear and polynomial equations for HMP 140A capacitive sensor using 5 salt solutions.

	Linear	2nd Order
b_0	0.226512	2.911321
b_1	1.023088	0.814217
b_2		0.00155423
R^2	0.9981	0.9995
s	1.4386	0.7890
Residual plots	clear pattern	Uniform distribution

The R^2 values for the linear and 2nd order polynomial calibration equations are 0.9981 and 0.9995, respectively. The s values for the linear and 2nd order polynomial calibration equations are 1.4386 and 0.7890, respectively. The 2nd order polynomial equations give the smallest estimated errors and listed as follows:

$y = 2.9113205 + 0.864217x + 0.0015542x^2$, $R^2 = 0.9995$, $s = 0.7890$
$(s_b = 0.63806 \; s_b = 0.02925 \; s_b = 0.000278$
$t = 74.5628 \; t = 29.543 \; t = 5.5872)$

The regression results for the 2nd order polynomial equations using different calibration points in different salt solutions are listed in Table 9. The results of R^2 values for 5, 7, 9 and 11 calibration points are similar. However, the calibration equation for 11 calibration points gives the smallest s value.

Table 9. Estimated parameters and evaluation criteria for the 2nd order polynomial equations for HMP 140A capacitive sensors using four different calibration points.

	Case 1 ($n_1 = 11$)	Case 2 ($n_2 = 9$)	Case 3 ($n_3 = 7$)	Case 4 ($n_4 = 5$)
b_0	3.479580	3.156891	2.871078	2.9113205
b_1	0.833274	0.844157	0.862302	0.8142171
b_2	0.00186718	0.00176878	0.00161775	0.00155423
R^2	0.9975	0.9992	0.9994	0.9995
s	0.6837	0.7127	0.7490	0.7890

3.1.3. Evaluation of Accuracy

The distribution between the number of saturated salt solutions and the estimated standard error for the calibration equations of two types of RH sensors is in Figure 6. For a resistance sensor, the s values of 7, 9, 11 calibration points are <0.8% RH. For a capacitance sensor, the s values for four saturated salt solutions are <0.8% RH. The accuracy of these calibration equations is <0.8% for both types of RH sensors. In terms a practical application [20,21], the calibration equation can be established using 7 salt solutions for a resistance sensor and 5 salt solutions for a capacitance sensor.

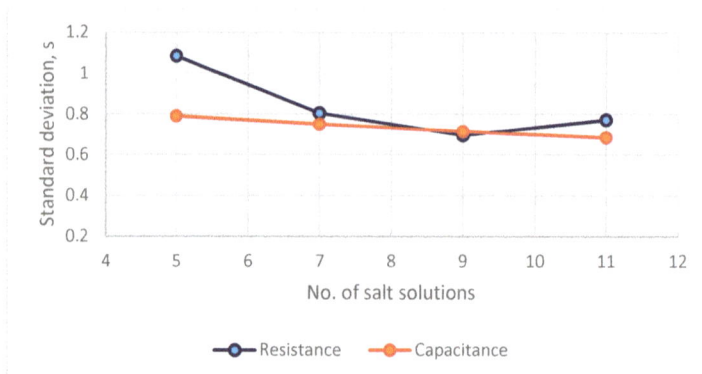

Figure 6. The distribution between numbers of saturated salt solutions and estimated standard errors of calibration equations of two types of RH sensors.

3.2. The Effect of the Precision of Calibration Points

3.2.1. The Measurement Uncertainty for the Two Humidity Sensors

The method that is used to calculate the measurement uncertainty is that of Lu and Chen [17]. Two Types "A" and "B" method are used to evaluate the measurement uncertainty. The Type A standard uncertainty is evaluated by statistical analysis of the experimental data. The Type B standard uncertainty is evaluated using other information that is related to the measurement.

The Type A standard uncertainty for the two types of humidity sensors used the uncertainty for the predicted values from the calibration equations. The Type B standard uncertainty for humidity sensors uses the reference standard, nonlinear and repeatability, resolution and temperature effect. The results for the Type B uncertainty analysis for resistive and capacitive sensors are respectively listed in Tables 10 and 11.

Table 10. The Type B uncertainty analysis for resistive humidity sensor.

Description	Estimate Value (%)	Standard Uncertainty u(x), (%)
Reference standard, U_{ref}		$N_1 = 11, u_{ref} = 0.3311$ $N_1 = 9, u_{ref} = 0.2983$ $N_1 = 7, u_{ref} = 0.3151$ $N_1 = 5, u_{ref} = 0.3084$
Non-linear and repeatability, U_{non}	±0.3	0.00866
Resolution, U_{res}	0.1	0.00290
The combined standard uncertainty of Type B = 0.1926		

Table 11. The Type B uncertainty analysis for capacitive humidity sensor.

Description	Estimate Value (%)	Standard Uncertainty u(x), (%)
Reference standard, U_{ref}		$N_1 = 11, u_{ref} = 0.3311$ $N_1 = 9, u_{ref} = 0.2983$ $N_1 = 7, u_{ref} = 0.3151$ $N_1 = 5, u_{ref} = 0.3084$
Nonlinear and repeatability, U_{non}	±0.1	0.0058
Resolution, U_{res}	±0.1	0.0029
Temperature effect, U_{temp}	±0.005	0.0043
The combined standard uncertainty of Type B = 0.1924		

The Type A standard uncertainty that are calculated using the predicted values for the 4th order polynomial equation for the resistive sensor and the 2nd order polynomial equation for a capacitive sensor are added to give a combined uncertainty using Equation (7). The combined uncertainty for three RH observations for the two humidity sensors using calibration equations that use different calibration points are in Figures 7 and 8.

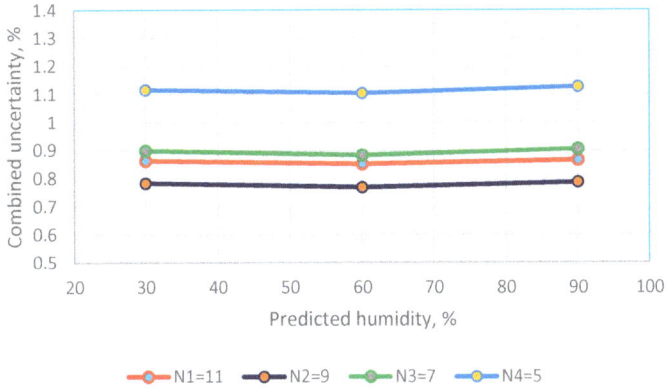

Figure 7. The distribution between numbers of saturated salt solutions and combined uncertainty of resistance RH sensors.

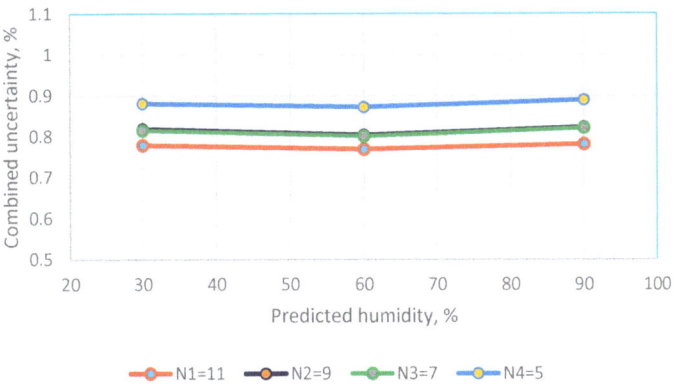

Figure 8. The distribution between numbers of saturated salt solutions and combined uncertainty of capacitance RH sensors.

3.2.2. The Precision of the Two Types of RH Sensors

The combined uncertainty is the criterion that is used to determine the precision of the sensors.

The values for the combined uncertainty for the resistive sensor at a RH of 30%, 60% and 90% are 0.8618%, 0.8506% and 0.8647% for the calibration equation that uses 11 calibration points, and 1.1155%, 1.1040% and 1.1271% for the calibration equation that uses 5 calibration points. The calibration equation that uses 9 calibration points gives the smallest u_c values. The combined uncertainty for 7, 9 and 11 calibration points is <1.0% RH.

The values for the combined uncertainty for a capacitive sensor at a RH of 30%, 60% and 90% are 0.7787%, 0.7690% and 0.7813% for the calibration equation that uses 11 calibration points and 0.8803%, 0.8717% and 0.8890% for the calibration equation that uses 5 calibration points. The combined uncertainty for 5, 7, 9 and 11 calibration points is <0.9% RH. In terms of practical applications, this performance is sufficient for industrial applications [20,21].

The accuracy and precision are 0.80% and 0.90% RH for a resistance RH sensor that uses 7 calibration points and 0.70% and 0.90% RH for a capacitance RH sensors that uses 5 calibration points.

3.3. Discussion

The number of calibration points that are required for sensors represents a compromise between the ideal number of calibration points and the time and cost of the calibration. The criterion that Betta [1] used to determine the optimal number of points used the ratio of the standard deviation of the regression coefficients (s_{bj}) to the established standard error of regression (s).

Accuracy and precision are the most important criteria for sensors so this study uses both values. Using statistical theory, the best calibration equation is determined using the *t*-value for the highest-order parameter and the residual plots. The estimated standard errors for the regression equation are then used to determine the accuracy of the sensors. The combined uncertainty considered the uncertainty of reference materials, the uncertainty for the predicted values and other B type sources. The combined uncertainties for the calibration equations for different numbers of calibration points using different saturated salt solutions are the criteria that are used to evaluate the precision of sensors.

Two types of electric RH sensors were calibrated in this study. Some calibration works, such as those for temperature and pressure sensors, are calibrated by an equal spacing of calibration points. The RH reference environments are maintained using different saturated salt solutions.

It is seen that the optimum number of calibration points that is required to calibrate a resistive humidity sensors involves 7 saturated salt solutions (LiCl, $MgCl_2$, K_2CO_3, NaBr, NaCl, KCl and K_2SO_4), so seven points are specified. Five saturated salt solutions (LiCl, $MgCl_2$, NaBr, NaCl and K_2SO_4) are specified for a capacitive humidity sensor. Considering factors that influence the choice of salts, such as price, toxicity and rules for disposal, the choice of these salt solutions is suitable.

The calibration equations key to measurement performance. This study determines that te 4th order polynomial equation is the adequate equation for the resistive humidity sensor and the 2nd order polynomial equation is the optimum equation for the capacitive humidity sensor. The accuracy of the calibration equations is 0.8% RH for a resistive humidity sensor that uses 7 calibration points and 0.7% RH for a capacitance humidity sensor that uses 5 calibration points. The precision is less than 1.0% RH for the resistive sensor and less than 0.9% RH for the capacitive sensor.

The method that is used in this study applicable to other sensors.

4. Conclusions

In this study, two types of electric RH sensors were used to illustrate the method for the specification of the optimum number of calibration points. The standard RH environments are maintained using different saturated salt solutions. The theory of regression analysis is applied. The best calibration equation is determined in terms of the *t*-value of the highest-order parameter and the residual plots. The estimated standard errors for the regression equation are the criteria that are used to determine the accuracy of sensors. The combined uncertainty involves the uncertainty for the reference materials, the uncertainty in the predicted values and other B type sources. The combined uncertainties for the calibration equations for different number of calibration points using different saturated salt solutions are the criteria that are used to evaluate the precision of the sensors.

The calibration equations are key to good measurement performance. This study determines that the 4th order polynomial equation is the adequate equation for the resistive humidity sensor and the 2nd order polynomial equation is the best equation for the capacitive humidity sensor. The accuracy of the calibration equations is 0.8% RH for a resistive humidity sensor that uses 7 calibration points and 0.7% RH for a capacitance humidity sensor using 5 calibration points. The precision is less than 1.0% RH for the resistive sensor and less than 0.9% RH for the capacitive sensor.

The method to determine the number of the calibration points used in this study is applicable to other sensors.

Supplementary Materials: The following are available online at http://www.mdpi.com/1424-8220/19/5/1213/s1. The residual plots for the calibration equations for different orders of polynomial equations for resistive humidity sensor using 5 saturated salt solutions (LiCl, $MgCl_2$, NaBr, NaCl and K_2SO_4). The residual plots for the calibration equations for different orders of polynomial equations for capacitance humidity sensor using 5 saturated salt solutions (LiCl, $MgCl_2$, NaBr, NaCl and K_2SO_4).

Author Contributions: H.-Y.C. drafted the proposal, executed the statistical analysis, interpreted the results and revised the manuscript. C.C. reviewed the proposal, performed some experiments, interpreted some results and criticized the manuscript and participated in its revision. All authors have read and approved the final manuscript.

Acknowledgments: The authors would like to thank the Ministry of Science and Technology of the Republic of China for financially supporting this research under Contract No. MOST -106-2313-B-005-006.

Conflicts of Interest: The authors declare no conflict of interest.

References

1. Betta, G.; Dell'Isola, M. Optimum choice of measurement points for sensor calibration. *Measurement* **1996**, *17*, 115–125. [CrossRef]
2. Betta, G.; Dell'Isola, M.; Frattolillo, A. Experimental design techniques for optimizing measurement chain calibration. *Measurement* **2001**, *30*, 115–127. [CrossRef]
3. Hajiyev, C. Determination of optimum measurement points via A-optimality criterion for the calibration of measurement apparatus. *Measurement* **2010**, *43*, 563–569. [CrossRef]
4. Hajiyev, C. Sensor Calibration Design Based on D-Optimality Criterion. *Metrol. Meas. Syst.* **2016**, *23*, 413–424. [CrossRef]
5. Khan, S.A.; Shabani, D.T.; Agarwala, A.K. Sensor calibration and compensation using artificial neural network. *ISA Trans.* **2003**, *42*, 337–352. [CrossRef]
6. Chen, C. Application of growth models to evaluate the microenvironmental conditions using tissue culture plantlets of *Phalaenopsis* Sogo Yukidian 'V3'. *Sci. Hortic.* **2015**, *191*, 25–30. [CrossRef]
7. Chen, H.; Chen, C. Use of modern regression analysis in liver volume prediction equation. *J. Med. Imaging Health Inform.* **2017**, *7*, 338–349. [CrossRef]
8. Wang, C.; Chen, C. Use of modern regression analysis in plant tissue culture. *Propag. Ornam. Plants* **2017**, *17*, 83–94.
9. Chen, C. Relationship between water activity and moisture content in floral honey. *Foods* **2019**, *8*, 30. [CrossRef]
10. Chen, C. Evaluation of resistance-temperature calibration equations for NTC thermistors. *Measurement* **2009**, *42*, 1103–1111. [CrossRef]
11. Chen, A.; Chen, C. Evaluation of piecewise polynomial equations for two types of thermocouples. *Sensors* **2013**, *13*, 17084–17097. [CrossRef] [PubMed]
12. ISO/IEC 98–3. *Uncertainty of Measurement—Part 3: Guide to the Expression of Uncertainty in Measurement*; ISO: Geneva, Switzerland, 2010.
13. National Aeronautics and Space Administration. *Measurement Uncertainty Analysis Principles and Methods, NASA Measurement Quality Assurance Handbook—Annex 3*; National Aeronautics and Space Administration: Washington, DC, USA, 2010.
14. Chen, C. Evaluation of measurement uncertainty for thermometers with calibration equations. *Accredit. Qual. Assur.* **2006**, *11*, 75–82. [CrossRef]
15. Myers, R.H. *Classical and Modern Regression with Applications*, 2nd ed.; Duxbury Press: Pacific Grove, CA, USA, 1990.
16. Weisberg, S. *Applied Linear Regression*, 4th ed.; Wiley: New York, NY, USA, 2013.
17. Lu, H.; Chen, C. Uncertainty evaluation of humidity sensors calibrated by saturated salt solutions. *Measurement* **2007**, *40*, 591–599. [CrossRef]
18. Greenspan, L. Humidity fixed points of binary saturated aqueous solutions. *J. Res. Natl. Bur. Stand.* **1977**, *81A*, 89–96. [CrossRef]

19. OMIL. *The Scale of Relative Humidity of Air Certified Against Saturated Salt Solutions*; OMIL R 121; Organization Internationale De Metrologie Legale: Paris, France, 1996.

20. Wernecke, R.; Wernecke, J. *Industrial Moisture and Humidity Measurement: A Practical Guide*; Wiley: Hoboken, NJ, USA, 2014.

21. Wiederhold, P.R. *Water Vapor Measurement*; Marcel Dekker, Inc.: New York, NY, USA, 1997.

22. Matko, V.; Đonlagić, D. Sensor for high-air-humidity measurement. *IEEE Trans. Instrum. Meas.* **1996**, *4*, 561–563. [CrossRef]

23. Matko, V. Next generation AT-cut quartz crystal sensing devices. *Sensors* **2011**, *5*, 4474–4482. [CrossRef] [PubMed]

24. Zheng, X.Y.; Fan, R.R.; Li, C.R.; Yang, X.Y.; Li, H.Z.; Lin, J.D.; Zhou, X.C.; Lv, R.X. A fast-response and highly linear humidity sensor based on quartz crystal microbalance. *Sens. Actuator B Chem.* **2019**, *283*, 659–665. [CrossRef]

25. Wexler, A.; Hasegawa, S. Relative humidity-temperature relationships of some saturated salt solutions in the temperature range 0° to 50° C. *J. Res. Natl. Bur. Stand.* **1954**, *53*, 19–26. [CrossRef]

26. Young, J. Humidity control in the laboratory using salt solutions—A review. *J. Chem. Technol. Biotechnol.* **1967**, *17*, 241–245. [CrossRef]

27. Lake, B.J.; Sonya, M.N.; Noor, S.M.; Freitag, H.P.; Michael, J.; McPhaden, M.J. *Calibration Procedures and Instrumental Accuracy Estimates of ATLAS Air Temperature and Relative Humidity Measurements*; NOAA Pacific Marine Environmental Laboratory: Seattle, WA, USA, 2003.

28. Wadsö, L.; Anderberg, A.; Åslund, I.; Söderman, O. An improved method to validate the relative humidity generation in sorption balances. *Eur. J. Pharm. Biopharm.* **2009**, *72*, 99–104. [CrossRef]

29. Duvernoy, J.; Gorman, J.; Groselj, D. A First Review of Calibration Devices Acceptable for Metrology Laboratory. 2015. Available online: https://www.wmo.int/pages/prog/www/IMOP/publications/IOM-94-TECO2006/4_Duvernoy_France.pdf (accessed on 11 December 2018).

30. Belhadj, O.; Rouchon, V. How to Check/Calibrate Your Hygrometer? *J. Paper Conserv.* **2015**, *16*, 40–41. [CrossRef]

31. Japan Mechanical Society. *The Measurement of Moisture and Humidity and Monitoring of Environment*; Japan Mechanical Society: Tokyo, Japan, 2011. (In Japanese)

32. Japan Industrial Standard Committee. *Testing Methods of Humidity*; JIS Z8866; JISC: Tokyo, Japan, 1998.

33. Centre Microcomputer Application. Relative Humidity Sensor 025I. Available online: http://www.cma-science.nl/resources/en/sensors_bt/d025i.pdf (accessed on 2 December 2018).

34. Delta Ohm Company. Calibration Instructions of Relative Humidity Sensors. 2012. Available online: http://www.deltaohm.com/ver2012/download/Humiset_M_uk.pdf (accessed on 10 December 2018).

35. Omega Company. Equilibrium Relative Humidity Saturated Salt Solutions. 2013. Available online: https://www.omega.com/temperature/z/pdf/z103.pdf (accessed on 11 December 2018).

36. TA Instruments. Humidity Fixed Points. 2016. Available online: http://www.tainstruments.com/pdf/literature/TN056.pdf (accessed on 10 December 2018).

37. Vaisala Ltd. Vaisala Humidity Calibrator HMK 15 User's Guide. 2017. Available online: www.vaisala.com/sites/default/files/documents/HMK15_User_Guide_in_English.pdf (accessed on 11 December 2018).

38. Ellison, S.; Williams, A. *Eurachem/CITAC Guide: Quantifying Uncertainty in Analytical Measurement*, 3rd ed.; Eurachem: Torino, Italy, 2012.

sensors

MDPI

Article

Humidity Sensors with Shielding Electrode Under Interdigitated Electrode

Hong Liu, Qi Wang, Wenjie Sheng, Xubo Wang, Kaidi Zhang, Lin Du and Jia Zhou *

ASIC and System State Key Lab, Department of Microelectronics, Fudan University, Shanghai 200433, China; 16210720074@fudan.edu.cn (H.L.); 18212020034@fudan.edu.cn (Q.W.); wsheng13@fudan.edu.cn (W.S.); xbwang16@fudan.edu.cn (X.W.); 15110720079@fudan.edu.cn (K.Z.); 17112020015@fudan.edu.cn (L.D.)
* Correspondence: jia.zhou@fudan.edu.cn; Tel.: +86-13818066203

Received: 14 December 2018; Accepted: 31 January 2019; Published: 6 February 2019

Abstract: Recently, humidity sensors have been investigated extensively due to their broad applications in chip fabrication, health care, agriculture, amongst others. We propose a capacitive humidity sensor with a shielding electrode under the interdigitated electrode (SIDE) based on polyimide (PI). Thanks to the shielding electrode, this humidity sensor combines the high sensitivity of parallel plate capacitive sensors and the fast response of interdigitated electrode capacitive sensors. We use COMSOL Multiphysics to design and optimize the SIDE structure. The experimental data show very good agreement with the simulation. The sensitivity of the SIDE sensor is $0.0063\% \pm 0.0002\%$ RH. Its response/recovery time is 20 s/22 s. The maximum capacitance drift under different relative humidity is 1.28% RH.

Keywords: humidity sensor; capacitive; PI; SIDE; IDE

1. Introduction

In addition to daily applications, such as air conditioners and humidifiers, humidity sensors are widely used in industrial process control, medical science, food production, agriculture, and meteorological monitoring [1–9]. In industry, the many manufacturing processes, such as semiconductor manufacturing and chemical gas purification, rely on precisely controlled humidity levels. In medical science, environmental humidity needs to be controlled during operations and pharmaceutical processing. In agriculture, humidity sensors are used for greenhouse air conditioning, plantation protection (dew prevention), soil moisture monitoring, and grain storage. Furthermore, in meteorological monitoring, weather bureaus and marine monitoring applications rely on accurate humidity sensing. For modern agriculture [10] and weather stations [11,12], accurate and fast measurement of humidity is becoming more and more important. Compared to existing infrared humidity sensors, electronic humidity sensors are cheaper, lighter, and smaller, which makes them more suitable for sensor networks to feed weather models. Nonetheless, high-precision fast-response sensors are important for many fields. For instance, fast and accurate humidity measurement are critical for eddy covariance systems [13]. Hence, electronic sensors have to become faster and more accurate.

Electronic humidity sensors can be divided into resistive and capacitive [14]. Resistive humidity sensors tend to have higher gain and are usually cheaper to manufacture than capacitive humidity sensors. However, these sensors do not respond well when operating at low relative humidity (about 10% RH) because they exhibit very poor conductivity in low relative humidity environments, making it difficult to measure the output response [15]. In contrast, capacitive humidity sensors have better linearity, accuracy, and higher thermal stability than resistive humidity sensors [16–19]. A capacitive humidity sensor responds to changes of humidity by changes of the relative dielectric constant of the sensing layer, e.g., polymer film, upon water vapor absorption. Therefore, it is possible to directly detect changes in capacitance to monitor changes in humidity. Unlike resistive humidity sensor, capacitive humidity sensors respond linearly with humidity, which simplifies the sensor readout.

Various materials can be used as humidity sensing materials, such as electrolyte [20], ceramics [21,22], porous inorganic material [23–26], and polymers [27–30]. In particular, polymers have been used as sensing materials for capacitive humidity sensors owing to their good dielectric properties arising from their microporous structure and measurable physical property changes due to water absorption. PI is among the most commonly used moisture sensing material [31] for its good mechanical strength, electrochemical stability, and flexibility [32]. It remains stable after long time exposure to the measurement environment. Furthermore, PI is a microporous material with imide groups that strongly bond water molecules, which makes the material dielectric constant very sensitive to humidity. Therefore, we used PI in the proposed capacitive sensor.

Capacitive humidity sensors have two basic structures: parallel plate (PP) capacitance (Figure 1a) and interdigital electrode (IDE) capacitance (Figure 1b).

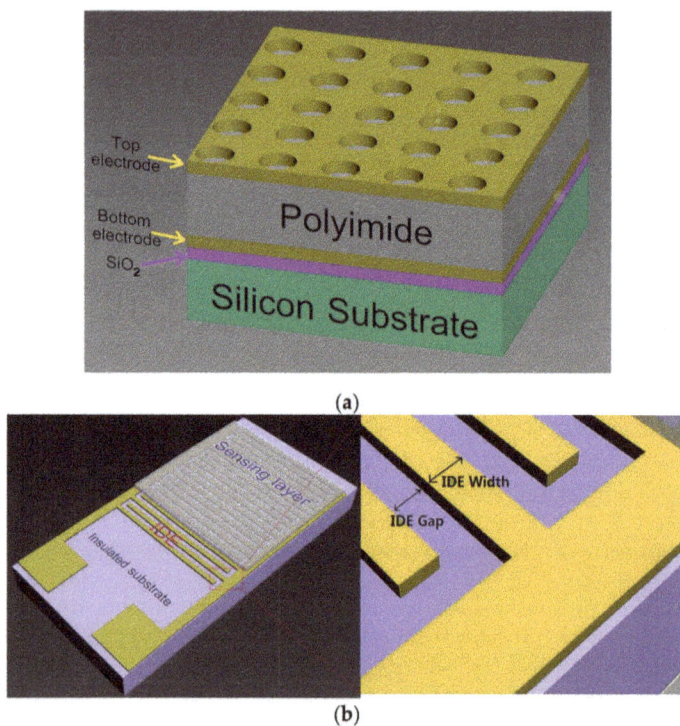

Figure 1. Structure diagram of parallel plate (PP) and interdigital electrode (IDE) sensors. (**a**) PP sensors composed of a solid substrate, two layers of parallel plate electrode, and a sensing material between them. (**b**) IDE sensors composed of an inert substrate, IDEs, and sensing material layer atop of the IDEs. A partial enlarged detail of IDE is shown on the right.

In PP sensors, the upper plate is perforated by an array of holes or parallel stripes to allow water molecules from the air to reach the sensing material underneath. Since the sensing area of the PP capacitor is sandwiched between two parallel plates, the change in the relative dielectric constant of the sensing material in the PP sensors affects the overall capacitance change. Unlike PP sensors, IDE sensors usually only affect the change in the upper capacitance of the IDEs, which makes them less sensitive than PP sensors. However, the exposed sensing area of the PP sensors is smaller than for IDE sensors, which causes a slower response than for IDEs.

The IDEs are fabricated on an inert solid or flexible substrate as parallel comb electrodes that overlap each other [6,33]. IDE sensors are easier to fabricate than PP ones. The sensitive area of the IDEs is typically a few square millimeters, and the electrode gap is a few microns. The sensitivity of this type of sensor increases with decreasing pitch [34]. The electric field strength above the IDEs decreases exponentially away from the electrode surface, and becomes one-thirtieth, or even lower, of the surface value [35] after a few microns. Therefore, in the case where the gap between the IDEs is several microns, a sensing layer only a few microns thick is enough. Thanks to this layer being completely exposed to the measurement environment, the IDE sensors are faster. However, in the IDEs, only half of the electric field lines pass through the sensing layer, and the other half of the electric field lines pass through the underlying substrate. Therefore, the IDE sensors will have only half or less sensitivity (depending on the relative dielectric constant of the substrate) compared to an equivalent PP sensor [36].

It is clear that there are advantages and disadvantages of these two types of sensors. There has been a significant effort to improve the sensor structures. For example, Zhao et al. used RIE (Reactive Ion Etching) and ICP (Inductively Couple Plasma) to etch sensing materials between parallel plates of the sensors to obtain a larger contact area with the tested environment to reduce response time from 35 s to 25 s [37], but this was still slower compared to typical equivalent IDEs.

Inspired by combining the advantages of PP and IDE structures, this paper proposes a novel IDE humidity sensor with a shielding electrode under the IDEs, namely, SIDE. On the SIDE, the capacitance of the lower half of the IDEs is shielded by an additional electrode underneath the IDEs, which effectively raises the relative capacitance change as it becomes exposed to moisture. Thus, a SIDE humidity sensor combines the high sensitivity of PP sensors and the fast response (20 s) as the IDE ones.

In this work, we first verified the feasibility of the SIDE structure in the simulation software. Secondly, the thickness of the sensing layer with different electrode gaps and the dielectric thickness between the shielding electrode and the IDEs were optimized regarding the sensitivity and response speed. The SIDE sensor with optimized parameters was fabricated. The sensitivity, response time, recovery time, and stability of the sensor were measured.

2. Simulation of SIDE

COMSOL Multiphysics®(Stockholm, Sweden) is applied to simulate the SIDE and IDE structure. Figure 2a shows the SIDE structure. The size of this sensor is 13 mm × 6 mm with a sensing area of 1.6 mm × 1 mm. The sensor consists of a 100 nm-thick shielding electrode, a 1 μm-thick silicon dioxide dielectric layer, a standard 100 nm IDE layer, and a PI film as the sensing layer. The finger length of the interdigitated electrode is 1 mm, with the width and the gap both being 5 μm. A total of 80 pairs of IDEs are used. A 5 μm-thick PI layer is utilized as the humidity sensing layer. Since the PI's relative dielectric constant increases linearly with humidity [38], we simulate variations of humidity by directly changing the relative dielectric constant of the PI. An IDE model with the same structural parameters as the SIDE one is implemented with the only difference being the absence of the shielding electrode.

Figure 2b shows the simulation results of the capacitance change rate ($\Delta C/C_0$) of SIDE and IDE under different relative dielectric constant of PI representing the humidity conditions. C_0 is the total capacitance when the relative dielectric constant of the sensing layer is 2.9. ΔC is the capacitance difference between any other relative dielectric constant of PI and 2.9. It can be seen that under the same conditions, $\Delta C/C_0$ of the SIDE structure, is about 4 times bigger than that of the IDE structure, which implies that the SIDE will have much higher sensitivity than IDE with the same parameters.

The effect of the thickness of the sensing film on $\Delta C_{max}/C_0$ is also simulated by COMSOL Multiphysics®(Stockholm, Sweden). We define that $\Delta C_{max}/C_0$ equals to $\Delta C/C_0$ with the relative dielectric constant of PI at 2.9 (C_0) and 3.7 (C_{max}), which indicates the sensitivity of the sensor.

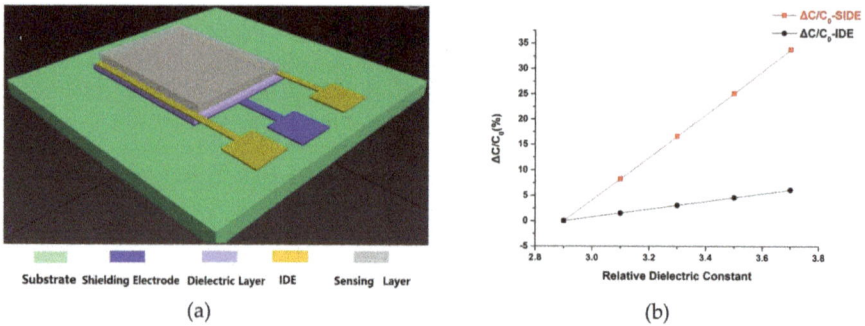

Figure 2. SIDE structure and simulation results. (**a**) 3D model of SIDE structure; (**b**) Comparison of the relative changes in capacitance of the SIDE (**red line**) and IDE (**black line**) structure according to numerical simulations.

Figure 3 shows that $\Delta C_{max}/C_0$ increases as the thickness of the sensing film increases, but flattens at higher thickness. To optimize the sensing film thickness, two facts should be taken into account. On the one hand, it is clear that when the sensing film thickness is equal to the gap between the IDEs (as those dashed lines in Figure 3), $\Delta C_{max}/C_0$ almost reaches saturated values. There is no significant increase of $\Delta C_{max}/C_0$ with thicker sensing film than the gap. On the other hand, the thickness of the sensing film also affects the speed of water molecules diffusing into the sensing film completely, which defines the sensor response and recovery time. Therefore, we select the optimized sensing film thickness as equal to the gap of the IDEs. Considering the laboratory conditions, we set the width and gap of the IDEs to 5 μm.

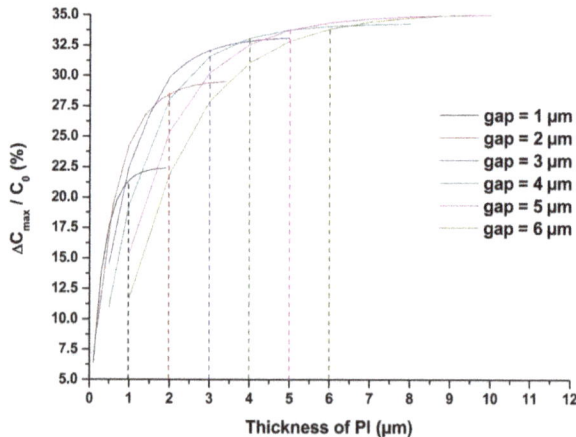

Figure 3. Influence of sensing film's thickness on sensor sensitivity. The vertical ordinate of the intersection of all the dashed lines and the solid curves represents the sensor's $\Delta C_{max}/C_0$ when the sensing film thickness is equal to the gap between the IDEs.

The effect of the spacing between the shielding electrode and the IDEs, i.e., the thickness of the silicon dioxide under the IDEs on the sensitivity in the SIDE structure is also studied.

Figure 4 shows that with the increasing thickness of the silicon dioxide layer, the $\Delta C_{max}/C_0$ increases first and then decreases, with an optimal value of the SiO_2 thickness of 1 μm.

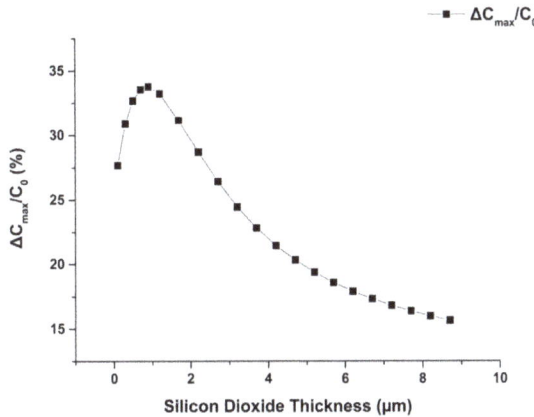

Figure 4. Influence of silicon dioxide thickness on the sensor sensitivity. For increasing silicon dioxide layer thickness, the full sensitivity increases first and then decreases past an optimal value.

There are several parameters of the optimized SIDE structure through the simulation: the gap of IDEs and spin-coated sensing film thickness are both 5 µm, and the thickness of the silicon dioxide layer is 1 µm. These parameters are used in the fabrication of the sensor.

3. Materials and Methods

The sensor is fabricated on a 3-inch silicon wafer according to the following steps: (a) A 2.5 µm-thick negative photoresist is patterned. (b) An e-beam-evaporated Ti/Au layer is deposited and selectively removed by a lift-off process to form the bottom shielding electrode. (c) A layer of 1 µm silicon dioxide is deposited by PECVD (Plasma Enhanced Chemical Vapor Deposition). (d) IDEs are fabricated on the silicon dioxide by the same sequence of lithography, e-beam evaporation, and lift-off. (e) A 5 µm-thick PI is spin-coated. Subsequently, the device is baked at 120 °C for 1 h, 180 °C for 1 h, and 250 °C for 6 h to cure the sensing layer. The completed sensor and cross-section of the SIDE structure under scanning electron microscope (SEM) are shown in Figure 5. The same IDE structure fabricated on the glass substrate without the shielding electrode is studied as the control experiment.

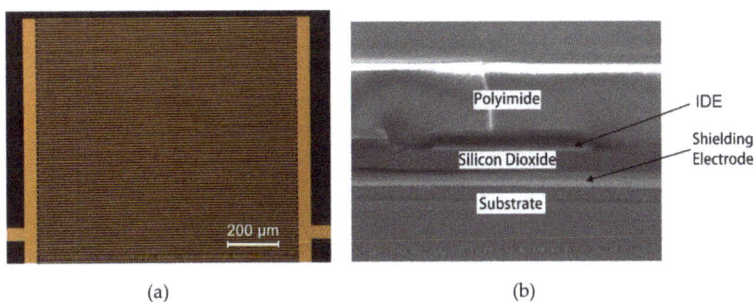

(a)   (b)

Figure 5. SIDE sensor picture under microscopy, and its cross-section image under SEM.

The setup for the humidity measurement is shown in Figure 6. The test is always carried out in an incubator. We build the simple incubator with heaters and semiconductor coolers inside. Each of them is controlled by an external PID (proportional integral derivative) controller to keep the temperature constant. In the incubator, we place a bottle of saturated salt solution and the sensor. The humidity is also monitored by a commercial humidity meter (Rotronic, HC2-S) at the same time and in the same incubator. The uncertainty of HC2-S is ±0.8% RH. The capacitance measurement uses an IC chip (SMARTEC's UTI03) and additional circuits. The commercial humidity sensor and the capacitance measurement circuit communicate with the computer using serial port simultaneously. The humidity and capacitance are recorded in parallel by the computer for later analysis.

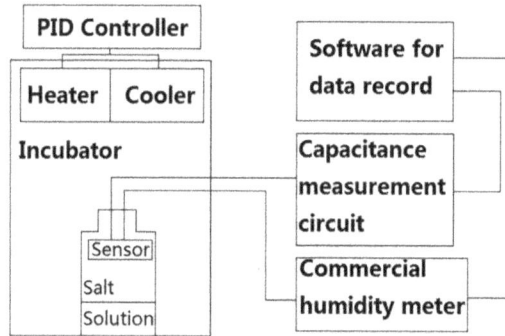

Figure 6. Block diagram of the measurement system consisting of an incubator, a measurement circuit and recording software.

The capacitance above the shielding electrode C_x can be directly measured using the circuit shown in Figure 7 without mixing the capacitance between the shielding electrode and IDEs C_{pn} (n = 1, 2). C_x is the sensing capacitance proportional to the humidity. C_{p1} and C_{p2} are the capacitances between the shielding electrode and the IDEs. C_f is the fixed capacitance of the IC chip. U_1 and U_2 are the potentials before the humidity sensor and after the IC chip that both can be measured. Therefore, C_x can be calculated using Equation (1).

$$C_x = -U_1/U_2 \cdot C_f \qquad (1)$$

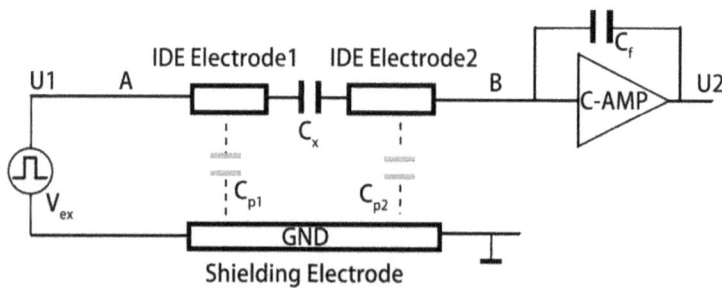

Figure 7. The working principle of the humidity capacitance measurement. The key point is to calculate the capacitance of C_x by measuring the induced charge generated at point B.

Before the test, each device is placed in an oven at 100 °C for 10 min to get rid of the effect of the previous measurement.

The sensitivity (S) can be expressed as Equation (2):

$$S = (\Delta C/C_0)/\Delta(\% \text{ RH}) \qquad (2)$$

where $\Delta C = C_1 - C_0$, C_0 is the capacitance measured at the RH, which is 23.7% ± 0.8%, and C_1 is the capacitance measured when the RH is 73.0% ± 0.8%. $\Delta(\% \text{ RH})$ is the difference between the relative humidity values when measuring C_1 and C_0.

The response and recovery dynamics are among the most important characteristics for evaluating the performance of humidity sensors. The response time for RH increase and the recovery time for RH decrease are usually defined for a sensor as the time taken to reach 90% of its total capacitance variation. The response and recovery curves are measured by exposing the SIDE sensor to alternate levels of humidity between 2.0% ± 0.8% and 77.0% ± 0.8% RH.

In order to evaluate the functioning of the humidity sensor over long periods of time, we measured the sensor's capacitance over the duration of 20 h at 25 °C with relative humidity levels of 25.7% ± 0.8%, 34.4% ± 0.8%, 45.0% ± 0.8%, 57.0% ± 0.8%, and 73.5% ± 0.8% RH.

4. Results and Discussion

A sensitivity test is carried out on the SIDE and IDE structure. Figure 8 shows the capacitance measured from SIDE and IDE at different levels of humidity, and their linear fits with R^2 of 0.996 and 0.991, respectively. The slopes of the line, i.e., S of SIDE and IDE are 0.0063 and 0.001,65, respectively. Taking the uncertainty of HC2-S into consideration, the S of SIDE and IDE are 0.0063 ± 0.0002 and 0.001,65 ± 0.000,05, respectively. Hence, the sensitivity of the SIDE structure is 3.82 times bigger than that of the IDE. These results show the significant improvement of sensitivity brought by the shielding electrode, that minimizes the large constant capacitance of the substrate. Indeed, whatever substrate the IDE is built on, the relative dielectric constant of the substrate is larger (e.g., Si is 11.9, glass is 10) or close to (e.g., flexible polymer films) the relative dielectric constant of PI (2.9–3.7). The experimental result and simulation data verify the effects of the shielding electrode and shows high agreement as well. It is clear that our proposed SIDE structure can provide an effective way to measure relative humidity more sensitively and accurately. Another advantage of the shielding electrode is that it can effectively suppress the external electromagnetic interference and reduce the noise in the measurement process.

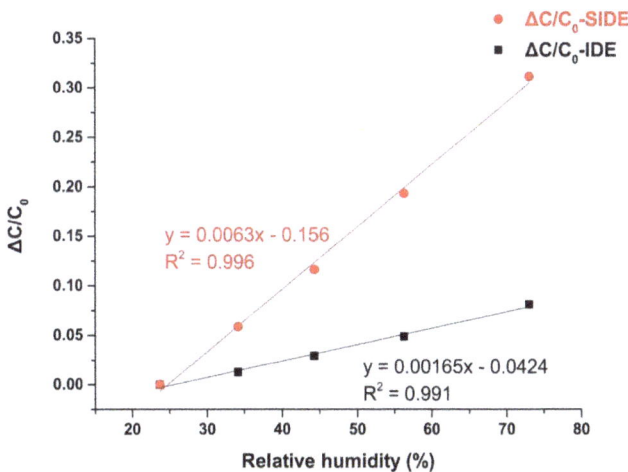

Figure 8. Experimental measurement of sensitivity of SIDE and IDE humidity sensors.

Figure 9 shows the responses of the SIDE sensor. The absorption curve represents the response of the sensor as a function of time, from an environment with low relative humidity to an environment with high relative humidity. The desorption curve represents the response of the sensor as a function of time, from an environment with high relative humidity to an environment with low relative humidity. The curve can switch to steady states rapidly after the RH level changes. Our sensor's response/recovery time is 20 s/22 s, which is comparable to 1 s/15 s for normal IDE reported in the literature [39], but a little worse. This is because in their work, the thickness of the sensing film is only 0.65 μm, while ours is 5 μm. If we scale down our sensors to reduce the IDE gap, the required sensing film thickness will also decrease, resulting in great improvement in response speed. Limited to laboratory conditions, we fabricated the sensor with 5 μm gap. However, our sensor's response/recovery time is still much better than 122 s for PP sensors [40].

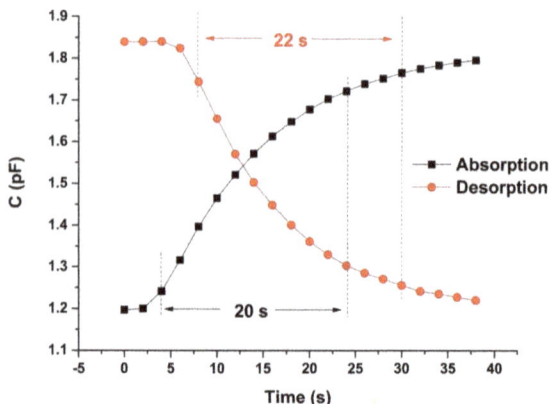

Figure 9. The response and recovery curves are measured by switching the SIDE sensor, alternately, between 2.0% ± 0.8% and 77.0% ± 0.8% RH. The response/recovery time is 20 s/22 s.

Figure 10 shows the stability characteristic of the SIDE sensor. The sensor is kept in the incubator for 20 h at 25.7% ± 0.8%, 34.4% ± 0.8%, 45.0% ± 0.8%, 57.0% ± 0.8%, and 73.5% ± 0.8% RH, respectively. The magnitude of the drift of sensor capacitance is converted into the apparent changes in relative humidity, D, which is calculated by

$$D = (C_{max} - C_{mean})/(C_0 \cdot S) \tag{3}$$

where C_{max} is the maximum measured capacitance after the sensor is exposed to different RH atmosphere, and C_{mean} is the average capacitance of all recorded values at a certain relative humidity, C_0 is the capacitance measured when the RH is 23.7% ± 0.8%. The maximum drift value (D) obtained from Figure 10 under different relative humidity was 1.28% RH. Thus, our sensor is able to achieve satisfactory stability from a practical standpoint, which makes it promising as a commercially available sensor.

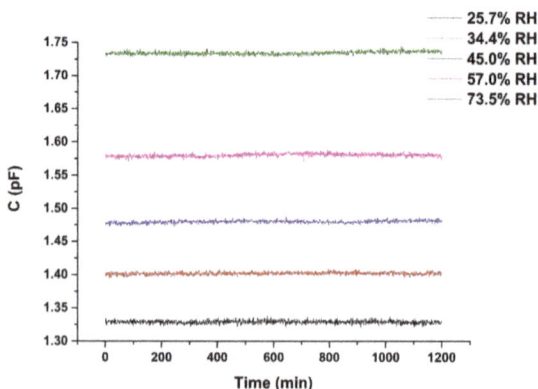

Figure 10. Stability of SIDE sensor. The sensor is kept in the incubator for 1200 min at 25.7% ± 0.8%, 34.4% ± 0.8%, 45.0% ± 0.8%, 57.0% ± 0.8%, and 73.5% ± 0.8% RH, respectively.

5. Conclusions

In summary, we propose a novel shielded interdigitated electrode structure for humidity sensing. We perform a comprehensive simulation of this structure to optimize the parameters for the sensor fabrication. In simulation and actual testing, we find that the sensitivity of the SIDE structure is much higher than that of the IDE structure because of the effect of the shielding electrode on the capacitance change rate. Since the surface structure of the SIDE structure is still the same as IDE, the SIDE sensor combines the high sensitivity of the parallel plate sensors and fast response of the IDE sensors. The sensitivity of SIDE is 0.0063% ± 0.0002% RH, and the response/recovery time is 20 s/22 s. The stability of the SIDE sensor was also characterized. The maximum drift value under different relative humidity is 1.28% RH.

Meanwhile, since the basic operating principle of many capacitive sensors is the same, the SIDE structure can even be applied to capacitive gas sensors, such as volatile organic compound (VOC) sensors which are used to monitor toxic gases. This shows that SIDE can replace IDE in various sensors that are more sensitive to the accuracy and response speed.

Author Contributions: Conceptualization, J.Z. and H.L.; methodology, H.L.; software, K.Z.; validation, Q.W., W.S. and L.D.; formal analysis, Q.W.; investigation, H.L.; resources, J.Z.; data curation, H.L.; writing—original draft preparation, H.L.; writing—review and editing, J.Z.; visualization, X.W.; supervision, J.Z.; project administration, J.Z.; funding acquisition, J.Z.

Acknowledgments: This work was supported by the National Natural Science Foundation of China (Grant No. 61874033), Science Foundation of Shanghai Municipal Government (Grant No.18ZR1402600) and the State Key Lab of ASIC and System, Fudan University with Grant No.2018MS003.

Conflicts of Interest: The authors declare no conflict of interest.

References

1. Tételin, A.; Pellet, C.; Laville, C.; N'Kaoua, G. Fast response humidity sensors for a medical microsystem. *Sensors Actuators B Chem.* **2003**, *91*, 211–218. [CrossRef]
2. Chen, Z.; Lu, C. Humidity Sensors: A Review of Materials and Mechanisms. *Sens. Lett.* **2005**, *3*, 274–295. [CrossRef]
3. Lee, C.W.; Lee, S.J.; Kim, M.; Kyung, Y.; Eom, K. Capacitive Humidity Sensor Tag Smart Refrigerator System using the Capacitive to Voltage Converter (CVC). *Int. J. Adv. Sci. Technol.* **2011**, *36*, 15–26.
4. Kolpakov, S.A.; Gordon, N.T.; Mou, C.; Zhou, K. Toward a new generation of photonic humidity sensors. *Sensors* **2014**, *14*, 3986–4013. [CrossRef] [PubMed]
5. Farahani, H.; Wagiran, R.; Hamidon, M.N. Humidity sensors principle, mechanism, and fabrication technologies: A comprehensive review. *Sensors* **2014**, *14*, 7881–7939. [CrossRef] [PubMed]
6. Pavinatto, F.J.; Paschoal, C.W.A.; Arias, A.C. Printed and flexible biosensor for antioxidants using interdigitated ink-jetted electrodes and gravure-deposited active layer. *Biosens. Bioelectron.* **2015**, *67*, 553–559. [CrossRef] [PubMed]
7. Lee, C.-Y.; Lee, G.-B. Humidity Sensors: A Review. *Sens. Lett.* **2005**, *3*, 1–15. [CrossRef]
8. Rittersma, Z.M. Recent achievements in miniaturised humidity sensors—A review of transduction techniques. *Sensors Actuators A Phys.* **2002**, *96*, 196–210. [CrossRef]
9. Willett, K.M.; Gillett, N.P.; Jones, P.D.; Thorne, P.W. Attribution of observed surface humidity changes to human influence. *Nature* **2007**, *449*, 710–712. [CrossRef] [PubMed]
10. Imam, S.A.; Choudhary, A.; Sachan, V.K. Design issues for wireless sensor networks and smart humidity sensors for precision agriculture: A review. In Proceedings of the 2015 International Conference on Soft Computing Techniques and Implementations (ICSCTI), Faridabad, India, 8–10 October 2015; pp. 181–187.
11. Chandana, L.S.; Sekhar, A.J.R. Weather Monitoring Using Wireless Sensor Networks based on IOT. *Int. J. Sci. Res. Sci. Technol.* **2018**, *4*, 525–531.
12. Yawut, C.; Kilaso, S. A Wireless Sensor Network for Weather and Disaster Alarm Systems. *Int. Conf. Inf. Electron. Eng.* **2011**, *6*, 155–159.
13. Baldocchi, D.D. Assessing the eddy covariance technique for evaluating carbon dioxide exchange rates of ecosystems: Past, present and future. *Glob. Chang. Biol.* **2003**, *9*, 479–492. [CrossRef]

14. Fenner, R.; Zdankiewicz, E. Micromachined Water Vapor Sensors: A Review of Sensing Technologies. *IEEE Sens. J.* **2001**, *1*, 309–317. [CrossRef]

15. Blank, T.A.; Eksperiandova, L.P.; Belikov, K.N. Recent trends of ceramic humidity sensors development: A review. *Sensors Actuators B Chem.* **2016**, *228*, 416–442. [CrossRef]

16. Dokmeci, M.; Najafi, K. A high-sensitivity polyimide capacitive relative humidity sensor for monitoring anodically bonded hermetic micropackages. *J. Microelectromech. Syst.* **2001**, *10*, 197–204. [CrossRef]

17. Gu, L.; Huang, Q.A.; Qin, M. A novel capacitive-type humidity sensor using CMOS fabrication technology. *Sensors Actuators B Chem.* **2004**, *99*, 491–498. [CrossRef]

18. Wagner, T.; Krotzky, S.; Weiß, A.; Sauerwald, T.; Kohl, C.D.; Roggenbuck, J.; Tiemann, M. A high temperature capacitive humidity sensor based on mesoporous silica. *Sensors* **2011**, *11*, 3135–3144. [CrossRef]

19. Lee, H.; Lee, S.; Jung, S.; Lee, J. Nano-grass polyimide-based humidity sensors. *Sensors Actuators B Chem.* **2011**, *154*, 2–8. [CrossRef]

20. Yang, M.-R.; Chen, K.-S. Humidity sensors using polyvinyl alcohol mixed with electrolytes. *Sensors Actuators B Chem.* **1998**, *49*, 240–247. [CrossRef]

21. Kim, Y.; Jung, B.; Lee, H.; Kim, H.; Lee, K.; Park, H. Capacitive humidity sensor design based on anodic aluminum oxide. *Sensors Actuators B Chem.* **2009**, *141*, 441–446. [CrossRef]

22. Feng, Z.S.; Chen, X.J.; Chen, J.J.; Hu, J. A novel humidity sensor based on alumina nanowire films. *J. Phys. D Appl. Phys.* **2012**, *45*, 225305. [CrossRef]

23. Tudorache, F.; Petrila, I. Effects of partial replacement of iron with tungsten on microstructure, electrical, magnetic and humidity properties of copper-zinc ferrite material. *J. Electron. Mater.* **2014**, *43*, 3522–3526. [CrossRef]

24. Tudorache, F.; Petrila, I.; Popa, K.; Catargiu, A.M. Electrical properties and humidity sensor characteristics of lead hydroxyapatite material. *Appl. Surf. Sci.* **2014**, *303*, 175–179. [CrossRef]

25. Tudorache, F.; Petrila, I.; Condurache-Bota, S.; Constantinescu, C.; Praisler, M. Humidity sensors applicative characteristics of granularized and porous Bi_2O_3 thin films prepared by oxygen plasma-assisted pulsed laser deposition. *Superlattices Microstruct.* **2015**, *77*, 276–285. [CrossRef]

26. Tudorache, F.; Petrila, I.; Slatineanu, T.; Dumitrescu, A.M.; Iordan, A.R.; Dobromir, M.; Palamaru, M.N. Humidity sensor characteristics and electrical properties of Ni–Zn–Dy ferrite material prepared using different chelating-fuel agents. *J. Mater. Sci. Mater. Electron.* **2016**, *27*, 272–278. [CrossRef]

27. Suzuki, T.; Tanner, P.; Thiel, D.V. O_2 plasma treated polyimide-based humidity sensors. *Analyst* **2002**, *127*, 1342–1346. [CrossRef] [PubMed]

28. Zampetti, E.; Pantalei, S.; Pecora, A.; Valletta, A.; Maiolo, L.; Minotti, A.; Macagnano, A.; Fortunato, G.; Bearzotti, A. Design and optimization of an ultra thin flexible capacitive humidity sensor. *Sensors Actuators B Chem.* **2009**, *143*, 302–307. [CrossRef]

29. Kim, J.H.; Hong, S.M.; Moon, B.M.; Kim, K. High-performance capacitive humidity sensor with novel electrode and polyimide layer based on MEMS technology. *Microsyst. Technol.* **2010**, *16*, 2017–2021. [CrossRef]

30. Liu, M.Q.; Wang, C.; Kim, N.Y. High-sensitivity and low-hysteresis porous mim-type capacitive humidity sensor using functional polymer mixed with TiO_2 microparticles. *Sensors* **2017**, *17*, 284. [CrossRef] [PubMed]

31. Wang, H.; Feng, C.-D.; Sun, S.-L.; Segre, C.U.; Stetter, J.R. Comparison of conductometric humidity-sensing polymers. *Sens. Actuators B Chem.* **1997**, *40*, 211–216. [CrossRef]

32. Fujita, S.; Kamei, Y. Electrical properties of polyimide with water absorption. In Proceedings of the 11th IEEE International Symposium on Electrets, Melbourne, VIC, Australia, 1–3 October 2002; pp. 275–278.

33. Olthuis, W.; Sprenkels, A.J.; Bomer, J.G.; Bergveld, P. Planar interdigitated electrolyte-conductivity sensors on an insulating substrate covered with Ta_2O_5. *Sensors Actuators B Chem.* **1997**, *43*, 211–216. [CrossRef]

34. Singh, K.V.; Bhura, D.K.; Nandamuri, G.; Whited, A.M.; Evans, D.; King, J.; Solanki, R. Nanoparticle-enhanced sensitivity of a nanogap-interdigitated electrode array impedimetric biosensor. *Langmuir* **2011**, *27*, 13931–13939. [CrossRef]

35. Schaur, S.; Jakoby, B. A numerically efficient method of modeling interdigitated electrodes for capacitive film sensing. *Procedia Eng.* **2011**, *25*, 431–434. [CrossRef]

36. Blue, R.; Uttamchandani, D. Chemicapacitors as a versatile platform for miniature gas and vapor sensors. *Meas. Sci. Technol.* **2017**, *28*, 22001–22024. [CrossRef]

37. Qiang, T.; Wang, C.; Liu, M.Q.; Adhikari, K.K.; Liang, J.G.; Wang, L.; Li, Y.; Wu, Y.M.; Yang, G.H.; Meng, F.Y.; et al. High-Performance porous MIM-type capacitive humidity sensor realized via inductive coupled plasma and reactive-Ion etching. *Sensors Actuators B Chem.* **2018**, *258*, 704–714. [CrossRef]

38. Schubert, P.J.; Nevin, J.H. A polyimide-based capacitive humidity sensor. *IEEE Trans. Electron Devices* **1985**, *32*, 1220–1223. [CrossRef]

39. Laville, C.; Delétage, J.Y.; Pellet, C. Humidity sensors for a pulmonary function diagnostic microsystem. *Sensors Actuators B Chem.* **2001**, *76*, 304–309. [CrossRef]

40. Kim, J.H.; Hong, S.M.; Lee, J.S.; Moon, B.M.; Kim, K. High sensitivity capacitive humidity sensor with a novel polyimide design fabricated by mems technology. In Proceedings of the 4th IEEE International Conference on Nano/Micro Engineered and Molecular Systems, NEMS 2009, Shenzhen, China, 5–8 January 2009; pp. 703–706.

sensors

MDPI

Article

A Fast Response—Recovery 3D Graphene Foam Humidity Sensor for User Interaction

Yu Yu [1,2], Yating Zhang [1,2,*], Lufan Jin [1,2], Zhiliang Chen [1,2], Yifan Li [1,2], Qingyan Li [1,2], Mingxuan Cao [1,2], Yongli Che [1,2], Junbo Yang [3] and Jianquan Yao [1,2]

[1] Department of Electrical and Electronic Engineering, South University of Science and Technology of China, Shenzhen 518055, China; yuyu1990@tju.edu.cn (Y.Y.); jlfking@tju.edu.cn (L.J.); chenzl@tju.edu.cn (Z.C.); yifanli@tju.edu.cn (Y.L.); liqingyan216@163.com (Q.L.); mingxuancao@tju.edu.cn (M.C.); cheyongli@tju.edu.cn (Y.C.); jqyao@tju.edu.cn (J.Y.)

[2] Key Laboratory of Opto-Electronics Information Technology, Ministry of Education, School of Precision Instruments and Opto-Electronics Engineering, Tianjin University, Tianjin 300072, China

[3] Center of Material Science, National University of Defense Technology, Changsha 410073, China; yangjunbo008@sohu.com

* Correspondence: yating@tju.edu.cn

Received: 9 October 2018; Accepted: 3 December 2018; Published: 8 December 2018

Abstract: Humidity sensors allow electronic devices to convert the water content in the environment into electronical signals by utilizing material properties and transduction techniques. Three-dimensional graphene foam (3DGF) can be exploited in humidity sensors due to its convenient features including low-mass density, large specific surface area, and excellent electrical. In this paper, 3DGF with super permeability to water enables humidity sensors to exhibit a broad relative humidities (RH) range, from 0% to 85.9%, with a fast response speed (response time: ~89 ms, recovery time: ~189 ms). To interpret the physical mechanism behind this, we constructed a 3DGF model decorated with water to calculate the energy structure and we carried out the CASTEP as implemented in Materials Studio 8.0. This can be ascribed to the donor effect, namely, the electronic donation of chemically adsorbed water molecules to the 3DGF surface. Furthermore, this device can be used for user interaction (UI) with unprecedented performance. These high performances support 3DGF as a promising material for humidity sensitive material.

Keywords: three-dimensional graphene foams; humidity sensor; fast response; user interaction

1. Introduction

Humidity sensors have aroused attention in many fields such as industry, agriculture, and environment [1,2], and medical devices [3]. Generally, they measure humidity through a variety of transduction techniques, including the use of resistive [4,5], capacitive [6], optical fiber [7], and field effect transistors [8,9]. There are also some high precision impedance-frequency transducers using quartz crystals which compensate temperature drift, and have fast response, as investigated by in Matko et al. [10]. In high air humidity measurement there is a problem with response time of the sensors in conventional methods. A solution for this problem is sensors for high air humidity measurement which use open capacitors with very low response time such as is described by Vojko et al. [11].

As an active material for absorbing water molecules, a series of sensing materials including polymers [12], metal oxides [8], carbon nanotubes [13,14], graphene dioxide [15,16], and composites [5,17] have been exploited in humidity sensors. For instance, Zhang et al. [12] described humidity sensors utilizing poly(N-vinyl-2-pyrrolidone) (PVP), poly(vinyl alcohol) (PVA), and hydroxyethyl cellulose (HEC). In particular, after using PVP, the humidity sensors exhibited response and recovery times between 11% and 95% relative humidity (RH) were about 37 s and 10 s, respectively. Wang et al. [8] applied a single SnO_2 nanowire (NW) to fabricate a humidity sensor, which exhibited a wide sensor RH range (5~85%), and the response and recovery times were 120~170 s and 20~60 s, respectively. Zhao et al. [15] investigated a humidity sensor based on multi-wall carbon nanotubes where the sensor testing range was about 11% to 97% RH, the response time was 45 s, and the recovery time was 15 s. Borini et al. [15] exploited graphene oxide in a humidity sensor and obtained an unprecedented response speed (~30 ms response and recovery times) in a range of 30% to 70% RH. Zhang et al. [5] utilized a graphene oxide (GO)/poly(diallyldimethylammonium chloride) (PDDA) nanocomposite film to fabricate a humidity sensor. The humidity sensor exhibited ultrahigh performance over a wide range of 11~97% RH, and the recovery time is 125 s at 11% RH. Thus, each sensing material has its own advantages and specific conditions of application. In addition, with large surface area to volume ratio, nanomaterials are attractive to fabricate humidity sensors with ultrahigh performance features including high sensitivity and fast response times.

Recently, graphene with three dimensional (3D) architectures, including foams, networks, and gels have been investigated [18–21]. These 3D graphene-based materials not only have the characteristics of graphene, but also have high specific surface area, low density, good mechanical strength and good conductivity [22]. Because of its wide accessibility, easy synthesis and solution processability, high chemical stability and strong adaptability [23,24], 3D graphene foam (3DGF) has attracted great interest in various sensing applications. Meanwhile, 3DGFs are efficient materials for biosensors and gas-sensing devices given their low-mass density, large surface area, good mechanical stability, and high electrical conductivity. Huang and coworkers [25] synthesised 3DGF/CuO nanoflower composites as single-chip independent 3D biosensors for the electrochemical detection of ascorbic acid with outstanding biosensing properties, such as an ultrahigh sensitivity of 2.06 mA mM^{-1} cm^{-2} to ascorbic acid at a 3 s response time. Besides that, Yavari et al. [24] used macroscopic 3DGF to fabricate gas detectors with high sensitivity. Generally, these electrical-type 3DGF sensors exhibit high sensitivity due to these properties including an ultrahigh surface area, and its electronic properties. It shows a strong dependence on surface absorbents (including gas molecules), which can change the carrier density of graphene [24]. Therefore, it is necessary to develop a new type of humidity sensor based on 3DGF by utilizing the unique structure and chemical characteristics and avoiding its shortcomings.

In this paper, we fabricate a humidity field effect transistor based on 3DGF and develop test equipment to measure the properties of the device. It exhibits a high performance over a broad RH range from 0% to 85.9%, with fast response and recovery times. To interpret the physical mechanism, we construct the 3DGF model decorated with water and apply CASTEP in the Materials Studio software to calculate the energy structure. Herrin, we explore the potential of 3D GF for portable, reliable and low cost humidity sensing applications in the future.

2. Materials and Methods

Utilizing a modified Hummers' method, [19–21] graphene oxide, denoted as GO, was synthesized from natural graphite powder by an oxidation reaction. GO ethanol solution (50 mL) with the concentration of 1 mg mL^{-1} was sealed in a 100 mL Teflon-lined autoclave which was then heated up to 180 °C and held for 12 h. Then the autoclave was cooled naturally to room temperature. The prepared ethanol intermediates were carefully removed from the autoclave by a slow and gradual solvent exchange with water. After the solvent exchange process was completed, the product filled with water was freeze-dried and then dried at 120 °C for 2 h in a vacuum oven. Finally, the sample was annealed at 450 °C in H_2/Ar (5/95, v/v) for 6 h. Finally, the sample was treated in a UV ozone system for 15 min to obtain the final 3DGF. The infrared spectrum of 3DGF was recorded on a Fourier transform infrared (FTIR) spectrophotometer using potassium bromide (KBr) pellets. Figure 1a shows the FTIR spectra of three-dimensional graphene foam (3DGF)) with water molecules (black line) and dry (red line) conditions. It can be seen that a broad peak at 3436 cm^{-1} corresponds to the vibration due to the stretching and bending of OH groups present in the water molecules adsorbed by 3DGF.

Thus, it was concluded that 3DGF exhibits strong hydrophilicity. Meanwhile, the absorption peaks at 565, 1163, and 1640 cm^{-1} correspond to the symmetric and antisymmetric stretching vibrations of C=O, C–O, and C–C groups for 3DGF, respectively. Figure 1b shows the surface morphologies of the 3DGF. Field emission scanning electron microscopy (SEM) images show clear, layered and interconnected three-dimensional uniform graphene sheets. It can be concluded that it forms a spongy porous network structure. [20]. The samples are cut into rectangular slabs (14 mm × 2 mm), and both sides are pasted by copper conductive adhesives on silicon substrates with a size of 14 mm × 14 mm for electrical contact.

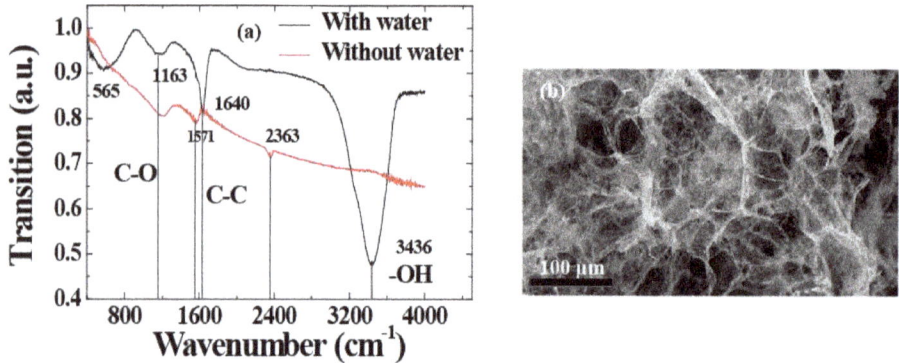

Figure 1. (a) FTIR spectra of 3D graphene with or without water molecule. (b) Field emission scanning electron microscopy (SEM) images of 3DGF.

For humidity sensors, chemical or physical reactions between water molecules and materials induce changes in channel current. External factors including the water concentration, temperature, and operating conditions will impact the performance of the device. For accurate measurements, as shown in Figure 2a, we used a closed box as an experimental chamber to control the humidity. In detail, the water concentrations were controlled by the ratio of saturated water vapor generated by a humidifier to high-purity nitrogen. We assure high quality humidity measurement in different ambient temperature operating conditions in climate chamber as shown in [26]. In order to measure the channel current flowing into the drain electrode (I_{DS}) [27–29], the source (with ground connection) and drain electrodes were connected with a Keithley 2400 apparatus (Tektronix China Ltd, Shanghai, China). The electrical measurements were also performed with this system, and the RH of the environment was measured by a commercial humidometer. Therefore, as described by Figure 2b, the output characteristics of the device were measured under dry and humid conditions. It shows that when the RH level was fixed to 100%, the channel current (I_{DS}) became lower than the conditions under drying. Meanwhile, the Dirac point shifted towards the positive direction. This donor effect [1] has been ascribed to the donation of electrons from the chemically adsorbed water molecules to the 3DGF. It can be concluded that the water molecules decorated in 3DGF will attract electrons and remain as holes, leading to p-type doping. Furthermore, water molecules decrease the charge mobility of 3D graphene, leading to lower currents. Through swelling or the 2D capillary effect [7,15,24], the dielectric constant will increase and the resistance decrease after adsorbing water molecules (confirmed using FTIR, as shown in Figure 1a). At the same time, the space charge polarization effect can be enhanced by adsorbing more water molecules, leading to the rapid diffusion of 3DGF and the formation of protons between hydroxyl groups. [6]. To investigate the mechanism, band energy of graphene decorating with water molecule was theoretically simulated by density functional theory (DFT) in the Material Studio 8.0 software (Neotrident Technology Ltd. Beijing, China). Simply speaking, graphene is simulated by plane wave program implemented in CASTEP. Considering the single and double supercells (2 × 1 × 1 allowing edge reconstruction) under GGA-PBE with 9 × 1 × 1 k-points Monkhorst-Pack point grid and 500 eV plane wave base truncation, the graphene is simulated by plane wave program with basis cutoff of 500 eV. The geometry was optimized until the total energy reached 2×10^{-5} eV/atom and the maximum force acting on each atom is less than 0.05 eV/Å. For the 3D graphene foam and 3D graphene foam adding water molecule calculations, the CASTEP plane wave code was used under GGA-PBE considering a Monkhorst−Pack grid with $9 \times 9 \times 1$ k-points and a plane wave basis cutoff of 500 eV; optimizing the geometry until the total energy reaches 2×10^{-5} eV/atom and the maximum force per atom exhibits values less than 0.05 eV/Å [30,31].

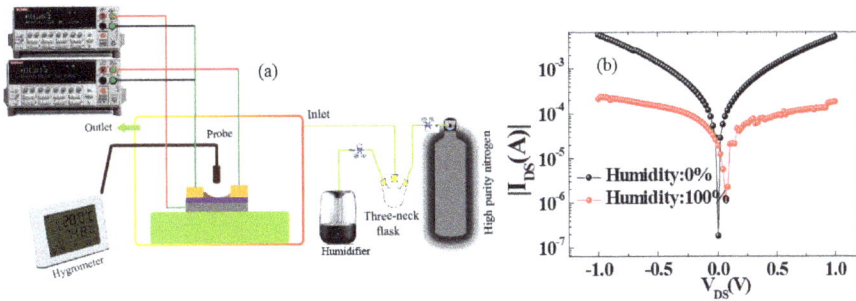

Figure 2. (a) Testing equipment used for the electrical characterization of 3DGF humidity sensors. (b) Output characteristic of the device decorated with or without water molecules.

3. Results and Discussion

Furthermore, the humidity-sensing performance of the 3DGF sensors exposed to different RH levels (0%, 10.0%, 19.9%, 30.3%, 44.5%, 51.4%, 57.1%, 60.3%, 66.4%, 70.5%, 75.2%, 80.2%, and 85.9% RH) are presented in Figure 3a. In a closed air-tight box, the humidity sensors were measured by different RH values ranging from 0 to 85.9%. It can be seen that as the RH level increased, the obtained channel currents of the sensor reduced monotonically. To consider the real-time response and recovery times of the devices, the time-dependent response and recovery curves of the device to 85.9% RH are plotted in Figure 3b. The time taken by a sensor to achieve 85% RH of the total channel current was defined as the response or recovery time. The response and recovery times of the sensor were approximately 89 ms and 189 ms, respectively. Additionally, our humidity sensors exhibited reproducibility and long-term stability. Professionally, the hysteresis value is a vital parameter for humidity sensors as it determines the maximum time lag between the response time (adsorption process) and recovery time (desorption process). With respect to the water content in the environment, the hysteresis effect is defined by the difference between the resistances. In particular, for a perfect humidity sensor, the hysteresis value should be as small as possible or can even be negligible.

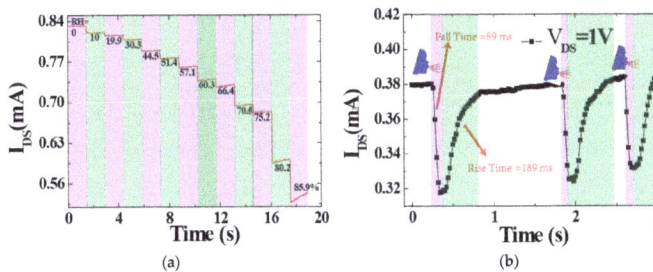

Figure 3. (a) Channel current response measurement of the 3DGF humidity sensor with varying different RH. (b) Response and recovery times of the device at 85% RH and the drain voltage was fixed at 1 V.

Table 1 compares the different characteristics of graphene-type humidity sensors including the response/recovery time, fabrication method, and sensitivity range. It was observed that the3DGF sensor exhibited broad sensitivity and rapid response and recovery rates.

Table 1. Comparison of different reported humidity sensors with graphene series materials.

Reference	Material	Sensing Range	Response/Recovery Time
Smith [30]	Graphene	1–96%	0.6 s/0.4 s
Ghosh [32]	Graphene	4–84%	180 s/180 s
Cai [33]	reduced graphene oxide (rGO)/graphene oxide (GO)/rGO	6.3–100%	1.9 s/3.9 s
Zhang [34]	Graphene oxide foam	36–92%	2 s/10 s
Trung [35]	rGO-polyurethane composites	10–70%	3.5 s/7 s
Leng [36]	GO/Nafion composite	11.3–97.3%	100–300 s/not shown
Bi [6]	GO	15–95%	10.5 s/41 s
Naik [37]	GO	30–95%	100 s/not shown
Yu [38]	GO/poly (sodium 4-styrenesulfonate) (PSS) composite	20–80%	60 s/50 s
Zhang [5]	rGO/poly(diallylimethyammonium chloride) PDDA composite	11–97%	108 s/94 s
Guo [39]	rGO	10–95%	50 s/3 s
This work	3DGF	0–85.9%	89 ms/189 ms

It can be seen that our devices showed good uniformity. Quantitatively, the effect of relative humidity on the device is depicted in Figure 4. Figure 4a describes the relationship between channel current and relative humidity. It can be seen that the relationship showed a decreasing trend with the increase in water humidity. This also showed that the channel current (I_{DS}) decreased more rapidly as relative humidity increased. To characterize the performance of the humidity sensor, the sensitivity (S) of the device was defined by Equation (1) [4,5,30,40]:

$$S = \frac{\left| I_{wet} - I_{dry} \right|}{I_{dry}\text{RH}} \times 100 \tag{1}$$

where I_{wet} and I_{dry} represent the channel current of the device under wet and dry conditions (RH = 0%), respectively. As shown in Figure 4b, the sensitivity increased rapidly as RH increased. Due to its perfect performance, including its ultrafast response/recovery rate, our humidity sensors can be used for breathing monitoring or for developing new user interfaces (UIs). Figure 3b presents the ability of a 3DGF sensor to monitor human breathing. In particular, during the user's speech, the ultrafast humidity sensor allowed the capture of fine features due to moisture modulation. Therefore, the 3DGF ultra-fast RH sensor can be used to identify different whistles, which can make use of low-cost and low-power sensors for user authentication.

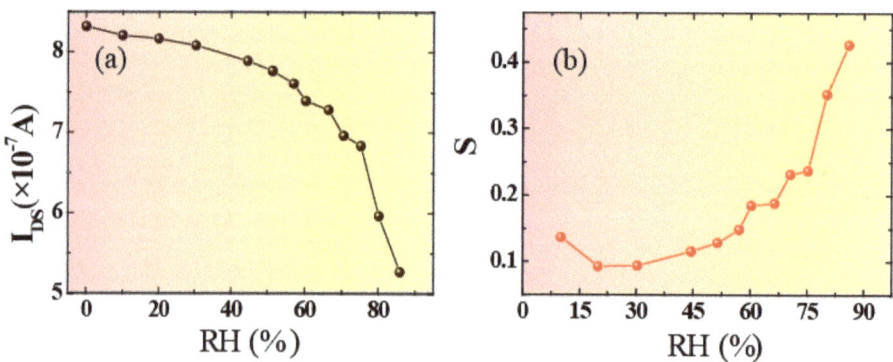

Figure 4. Relative humidity effect on the device performance. (**a**) Channel currents (I_{DS}) with the relationship of RH (**b**) The variation in sensitivity of the device for different RH values.

A schematic model of humidity sensing at a 3DGF film is shown in Figure 5a. To investigate its mechanism, the band energy of graphene decorated with water molecules was theoretically simulated by density functional theory (DFT) in the Material Studio 8.0 software. As shown in Figure 5b, conductivity and valence are at K Brillouin point, which makes the material a direct bandgap semiconductor. The direct band gap at the K point was ~0.172 eV, as shown in Figure 5c. This can be ascribed to the donor effect [3] attributed to the donation of electrons from the chemically adsorbed water molecules to the 3DGF surface. The water molecules decorated in 3DGF will attract electrons. Simply, water molecules open the band gap of 3DGF. Meanwhile, electron density will decrease and the conduction level will rise, leading to the formation of band energy.

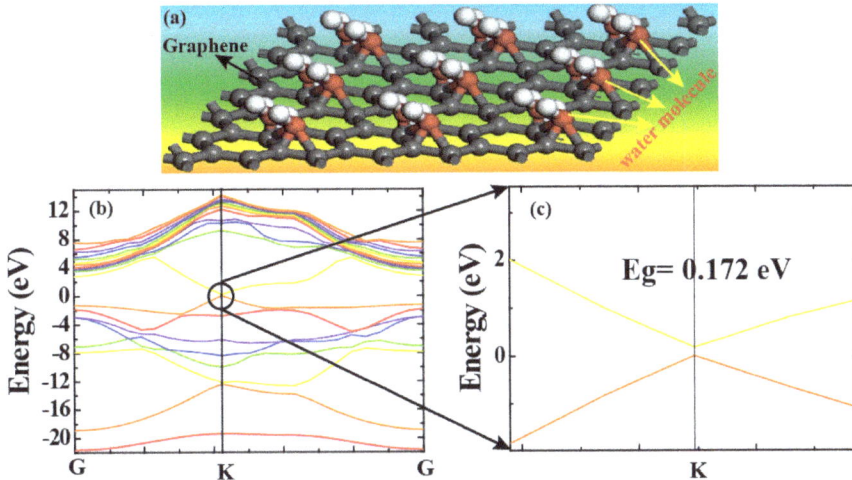

Figure 5. (**a**) The bonding mechanism between the graphene and water molecules. (**b**) The electronic band structure of graphene decorated with water. (**c**) The energy gap at the K point location.

4. Conclusions

In summary, a three-dimensional graphene foam (3DGF) exhibiting super permeability to water was exploited in humidity sensors, enabling a humidity sensor with a broad range of % RH values and unprecedented response speed (response time: ~89 ms, recovery time: ~189 ms). The ultra-fast response speed of these sensors enables us to observe the regulation of moisture in a user's breath. We constructed the 3DGF model decorated with water molecules theoretically and conducted the CASTEP as implemented in Materials Studio to calculate the energy structure. This allows sensors to be used in a variety of applications, such as humidity sensing, which we have experimentally verified with a cheap and easily available identification system. In addition, for different 3D materials, such as 3D transition metal dihalogenated hydrocarbons, ultra-thin nanoporous membranes for sensing applications can be realized in the interaction with different vapors and gases, which can be explored.

Author Contributions: Conceptualization, Y.Y. and Y.Z.; methodology, Y.Z.; formal analysis, Y.Y., Z.C., Y.L., L.J., Q.L., Y.C., and M.C; data curation, Y.Y.; writing—original draft preparation, Y.Y.; writing—review and editing, Y.Y.; supervision, Y.Z. and J.Y. (Junbo Yang); project administration, Y.Z. and J.Y. (Jianquan Yao); funding acquisition, J.Y. (Jianquan Yao).

Funding: This work was supported by the National Natural Science Foundation of China (Nos. 61675147, 61605141 and 61735010), Basic Research Program of Shenzhen (JCYJ20170412154447469) and Wenzhou City Governmetal Public Industrial Technology Project (G20160014).

Acknowledgments: We thank Yongshen Chen group in Nankai University, which he provides three dimensional graphene foam.

Conflicts of Interest: The authors declare no conflict of interest.

References

1. Traversa, E. Ceramic sensors for humidity detection: The state-of-the-art and future developments. *Sens. Actuators B Chem.* **1995**, *23*, 135–156. [CrossRef]
2. Chu, J.; Peng, X.; Feng, P.; Sheng, Y.; Zhang, J. Study of humidity sensors based on nanostructured carbon films produced by physical vapor deposition. *Sens. Actuators B Chem.* **2013**, *178*, 508–513. [CrossRef]
3. Chen, Z.; Lu, C. Humidity Sensors: A Review of Materials and Mechanisms. *Sens. Lett.* **2005**, *3*, 274–295. [CrossRef]
4. Mogera, U.; Sagade, A.A.; George, S.J.; Kulkarni, G.U. Ultrafast response humidity sensor using supramolecular nanofibre and its application in monitoring breath humidity and flow. *Sci. Rep.* **2014**, *4*, 4103. [CrossRef] [PubMed]
5. Zhang, D.; Tong, J.; Xia, B. Humidity-sensing properties of chemically reduced graphene oxide/polymer nanocomposite film sensor based on layer-by-layer nano self-assembly. *Sens. Actuators B Chem* **2014**, *197*, 66–72. [CrossRef]
6. Bi, H.; Yin, K.; Xie, X.; Ji, J.; Wan, S.; Sun, L.; Terrones, M.; Dresselhaus, M.S. Ultrahigh humidity sensitivity of graphene oxide. *Sci. Rep.* **2013**, *3*, 2714. [CrossRef] [PubMed]
7. Wong, W.C.; Chan, C.C.; Chen, L.H.; Li, T.; Lee, K.X.; Leong, K.C. Polyvinyl alcohol coated photonic crystal optical fiber sensor for humidity measurement. *Sens. Actuators B Chem.* **2012**, *174*, 563–569. [CrossRef]
8. Kuang, Q.; Lao, C.; Wang, Z.L.; Xie, Z.; Zheng, L. High-Sensitivity Humidity Sensor Based on a Single SnO$_2$ Nanowire. *J. Am. Chem. Soc.* **2007**, *129*, 6070–6071. [CrossRef]
9. Hu, P.; Zhang, J.; Li, L.; Wang, Z.; O'Neill, W.; Estrela, P. Carbon nanostructure-based field-effect transistors for label-free chemical/biological sensors. *Sensors* **2010**, *10*, 5133–5159. [CrossRef]
10. Vojko, M. Next generation AT-cut quartz crystal sensing devices. *Sensors* **2011**, *11*, 4474–4482.
11. Matko, V.; Donlagic, D. Sensor for high-air-humidity measurement. *Sens. Actuators A Phys.* **1997**, *61*, 331–334. [CrossRef]
12. Fei, T.; Zhao, H.; Jiang, K.; Zhou, X.; Zhang, T. Polymeric humidity sensors with nonlinear response: Properties and mechanism investigation. *J. Appl. Polym. Sci.* **2013**, *130*, 2056–2061. [CrossRef]
13. Chen, W.P.; Zhao, Z.G.; Liu, X.W.; Zhang, Z.X.; Suo, C.G. A Capacitive Humidity Sensor Based on Multi-Wall Carbon Nanotubes (MWCNTs). *Sensors* **2009**, *9*, 7431–7444. [CrossRef] [PubMed]
14. Han, J.-W.; Kim, B.; Li, J.; Meyyappan, M. A carbon nanotube based ammonia sensor on cellulose paper. *RSC Adv.* **2014**, *4*, 549–553. [CrossRef]
15. Borini, S.; White, R.; Wei, D.; Astley, M.; Haque, S.; Spigone, E.; Harris, N.; Kivioja, J.; Ryhänen, T. Ultrafast Graphene Oxide Humidity Sensors. *ACS Nano* **2013**, *7*, 11166–11173. [CrossRef] [PubMed]
16. Zhao, X.; Long, Y.; Yang, T.; Li, J.; Zhu, H. Simultaneous High Sensitivity Sensing of Temperature and Humidity with Graphene Woven Fabrics. *ACS Appl. Mater. Interfaces* **2017**, *9*, 30171–30176. [CrossRef] [PubMed]
17. Zhang, D.; Tong, J.; Xia, B.; Xue, Q. Ultrahigh performance humidity sensor based on layer-by-layer self-assembly of graphene oxide/polyelectrolyte nanocomposite film. *Sens. Actuators B Chem.* **2014**, *203*, 263–270. [CrossRef]
18. Ma, Y.; Chen, Y. Three-dimensional graphene networks: Synthesis, properties and applications. *Natl. Sci. Rev.* **2015**, *2*, 40–53. [CrossRef]
19. Wu, Y.; Yi, N.; Huang, L.; Zhang, T.; Fang, S.; Chang, H.; Li, N.; Oh, J.; Lee, J.A.; Kozlov, M.; et al. Three-dimensionally bonded spongy graphene material with super compressive elasticity and near-zero Poisson's ratio. *Nat. Commun.* **2015**, *6*, 6141. [CrossRef]
20. Zhang, T.; Chang, H.; Wu, Y.; Xiao, P.; Yi, N.; Lu, Y.; Ma, Y.; Huang, Y.; Zhao, K.; Yan, X.-Q.; et al. Macroscopic and direct light propulsion of bulk graphene material. *Nat. Photonics* **2015**, *9*, 471–476. [CrossRef]
21. Chang, H.; Qin, J.; Xiao, P.; Yang, Y.; Zhang, T.; Ma, Y.; Huang, Y.; Chen, Y. Highly Reversible and Recyclable Absorption under Both Hydrophobic and Hydrophilic Conditions using a Reduced Bulk Graphene Oxide Material. *Adv. Mater.* **2016**, *28*, 3504–3509. [CrossRef]
22. Cao, X.; Yin, Z.; Zhang, H. Three-dimensional graphene materials: Preparation, structures and application in supercapacitors. *Energy Environ. Sci.* **2014**, *7*, 1850–1865. [CrossRef]

23. Chabot, V.; Higgins, D.; Yu, A.; Xiao, X.; Chen, Z.; Zhang, J. A review of graphene and graphene oxide sponge: Material synthesis and applications to energy and the environment. *Energy Environ. Sci.* **2014**, *7*, 1564–1596. [CrossRef]

24. Yavari, F.; Chen, Z.; Thomas, A.V.; Ren, W.; Cheng, H.M.; Koratkar, N. High sensitivity gas detection using a macroscopic three-dimensional graphene foam network. *Sci. Rep.* **2011**, *1*, 166. [CrossRef]

25. Ma, Y.; Zhao, M.; Cai, B.; Wang, W.; Ye, Z.; Huang, J. 3D graphene foams decorated by CuO nanoflowers for ultrasensitive ascorbic acid detection. *Biosens. Bioelectron.* **2014**, *59*, 384–388. [CrossRef] [PubMed]

26. Brezovec, B.; Matko, V. Software and Equipment for Remote Testing of Sensors. *Sensors* **2007**, *7*, 1306–1316. [CrossRef]

27. Yu, Y.; Zhang, Y.; Zhang, Z.; Zhang, H.; Song, X.; Cao, M.; Che, Y.; Dai, H.; Yang, J.; Wang, J.; et al. Broadband Phototransistor Based on $CH_3NH_3PbI_3$ Perovskite and PbSe Quantum Dot Heterojunction. *J. Phys. Chem. Lett.* **2017**, *8*, 445–451. [CrossRef]

28. Yu, Y.; Zhang, Y.; Song, X.; Zhang, H.; Cao, M.; Che, Y.; Dai, H.; Yang, J.; Zhang, H.; Yao, J. High Performances for Solution-Pocessed 0D-0D Heterojunction Phototransistors. *Adv. Opt. Mater.* **2017**, *5*, 1700565. [CrossRef]

29. Yu, Y.; Zhang, Y.; Song, X.; Zhang, H.; Cao, M.; Che, Y.; Dai, H.; Yang, J.; Zhang, H.; Yao, J. PbS-Decorated WS_2 Phototransistors with Fast Response. *ACS Photonics* **2017**, *4*, 950–956. [CrossRef]

30. Smith, A.D.; Elgammal, K.; Niklaus, F.; Delin, A.; Fischer, A.C.; Vaziri, S.; Forsberg, F.; Rasander, M.; Hugosson, H.; Bergqvist, L.; et al. Resistive graphene humidity sensors with rapid and direct electrical readout. *Nanoscale* **2015**, *7*, 19099–19109. [CrossRef]

31. Gutierrez, H.R.; Perea-Lopez, N.; Elias, A.L.; Berkdemir, A.; Wang, B.; Lv, R.; Lopez-Urias, F.; Crespi, V.H.; Terrones, H.; Terrones, M. Extraordinary room-temperature photoluminescence in triangular WS_2 monolayers. *Nano Lett.* **2013**, *13*, 3447–3454. [CrossRef] [PubMed]

32. Ghosh, A.; Late, D.J.; Panchakarla, L.S.; Govindaraj, A.; Rao, C.N.R. NO_2 and humidity sensing characteristics of few-layer graphenes. *J. Exp. Nanosci.* **2009**, *4*, 313–322. [CrossRef]

33. Cai, J.; Lv, C.; Aoyagi, E.; Ogawa, S.; Watanabe, A. Laser Direct Writing of a High-Performance All-Graphene Humidity Sensor Working in a Novel Sensing Mode for Portable Electronics. *ACS Appl. Mater. Interfaces* **2018**, *10*, 23987–23996. [CrossRef] [PubMed]

34. Zhang, K.-L.; Hou, Z.-L.; Zhang, B.-X.; Zhao, Q.-L. Highly sensitive humidity sensor based on graphene oxide foam. *Appl. Phys. Lett.* **2017**, *111*, 153101. [CrossRef]

35. Trung, T.Q.; Duy, L.T.; Ramasundaram, S.; Lee, N.-E. Transparent, stretchable, and rapid-response humidity sensor for body-attachable wearable electronics. *Nano Res.* **2017**, *10*, 2021–2033. [CrossRef]

36. Leng, X.; Luo, D.; Xu, Z.; Wang, F. Modified graphene oxide/Nafion composite humidity sensor and its linear response to the relative humidity. *Sens. Actuators B Chem.* **2018**, *257*, 372–381. [CrossRef]

37. Naik, G.; Krishnaswamy, S. Room-Temperature Humidity Sensing Using Graphene Oxide Thin Films. *Graphene* **2016**, *5*, 1–13. [CrossRef]

38. Yu, H.W.; Kim, H.K.; Kim, T.; Bae, K.M.; Seo, S.M.; Kim, J.M.; Kang, T.J.; Kim, Y.H. Self-powered humidity sensor based on graphene oxide composite film intercalated by poly(sodium 4-styrenesulfonate). *ACS Appl. Mater. Interfaces* **2014**, *6*, 8320–8326. [CrossRef]

39. Guo, L.; Jiang, H.B.; Shao, R.Q.; Zhang, Y.L.; Xie, S.Y.; Wang, J.N.; Li, X.B.; Jiang, F.; Chen, Q.D.; Zhang, T. Two-beam-laser interference mediated reduction, patterning and nanostructuring of graphene oxide for the production of a flexible humidity sensing device. *Carbon* **2012**, *50*, 1667–1673. [CrossRef]

40. Zhu, Z.-T.; Mason, J.T.; Dieckmann, R.; Malliaras, G.G. Humidity sensors based on pentacene thin-film transistors. *Appl. Phys. Lett.* **2002**, *81*, 4643–4645. [CrossRef]

sensors

MDPI

Article

Design and Implementation of an Infrared Radiant Source for Humidity Testing

Hong Zhang *, Chuansheng Wang, Xiaorui Li, Boyan Sun and Dong Jiang

School of Computer Science and Technology, Harbin University of Science and Technology, 52 Xuefu Road, Harbin 150080, China; wangchuansheng994@163.com (C.W.); 13682088813@163.com (X.L.); 15754509280@163.com (B.S.); wdyu2004@163.com (D.J.)
* Correspondence: zhangh@hrbust.edu.cn; Tel.: +86-177-6655-5090 or +86-138-3609-9065

Received: 26 June 2018; Accepted: 11 September 2018; Published: 13 September 2018

Abstract: A novel way to measure humidity through testing the emissivity of an area radiant source is presented in this paper. The method can be applied in the environment at near room temperature (5~95 °C) across the relative humidity (RH) range of 20~90% RH. The source, with a grooved radiant surface, works in the far infrared wavelength band of 8~12 μm. The Monte-Carlo model for thermal radiation was set up to analyze the V-grooved radiant surface. Heat pipe technology is used to maintain an isothermal radiant surface. The fuzzy-PID control method was adopted to solve the problems of intense heat inertia and being easily interfered by the environment. This enabled the system to be used robustly across a large temperature range with high precision. The experimental results tested with a scanning radiant thermometer showed that the radiant source can provide a uniform thermal radiation capable of satisfying the requirements of humidity testing. The calibration method for the radiant source for humidity was explored, which is available for testing humidity.

Keywords: infrared radiant source; Monte Carlo method; emissivity; calibration; humidity

---

1. Introduction

Compared to traditional humidity measurement methods, innovative electronic testing methods involving humidity sensors such as hygristors and humicaps are the current research direction. Even though electric methods have fast responses, they often lack stability and their accuracy is improved little. Widely available commercial humidity sensors composed of humicaps use embedded microprocessors, such as the DHT11 with ±5% RH precision, and 1% RH resolution which are convenient to use. Because humidity is often mingled together with temperature, the precision and the humidity measurement range are easily affected by the temperature. Humidity, which reflects the degree of dryness of the atmosphere, is an important variable that is extensively tested in agriculture, industry, hospital and warehouse. Some sensors with new materials possessing resistive and capacitive features have been explored, which include a sulfonated polycarbonate resistive humidity sensor [1], polyimide-based capacitive humidity sensor [2], a high-performance capacitive humidity sensor with novel electrode and polyimide layer capacitive humidity sensors [3], and some sensors with improved sensing properties whose response and recovery times are 14.5 s and 34.27 s, respectively, for humidity levels between 33% RH and 95% RH at 102 Hz. [4]. These kinds of humidity sensors often have long response times. Most studies focus on material and processing innovations to increase the humidity testing precision. There have been few breakthroughs in hygristors because these are easily interfered by the environment. Novel ways using new effects such as optical properties present approaches to measure humidity [5]. Because humidity is a factor that affects the radiation measurement, humidity can be measured indirectly by testing radiation changes [6,7].

As a standard radiant source, a blackbody is usually adopted for calibrating infrared instruments such as pyrometers and radiant thermometers. Industrial blackbodies ranging from −50~2500 °C have been developed by the National Research Council (NRC) of Canada in order to calibrate optical testing devices [8]. A high-spatial-resolution multi-spectral imager (ASTER) on the first platform (Terra) of NASA's Earth Observing System requires a blackbody radiant source on a satellite for calibration purposes [9]. Traditional blackbody cavities evaluated by the Bedford methods [10] usually possess symmetrical shapes with small apertures, which makes them suitable for the high temperature range case, but not for the case of near room temperature range measurements. Minimum Resolvable Temperature Difference (MRTD) sensitivity requires that a radiant source working in the far infrared scope should be an area source which can provide a stable radiant flux with high uniformity. As to environmental humidity measurement, according to MRTD, an area radiant source ranging from 5~95 °C in the wavelength bands of 8~12 μm, is required. Under a certain temperature, a source should emit a stable radiation which is monitored by a radiometer. In order to enhance the effective emissivity of a radiant surface, its surface is often processed into grooves or mini holes. The radiant surface of the source possesses concentric V-grooves which can increase the effective emissivity.

Among statistical evaluation methods, Monte Carlo methods have been widely applied in optical radiometry and blackbody cavity analysis [11]. Monte Carlo methods possess advantages which are greater than exactitude methods in complex radiant characteristic analysis. Therefore, the Monte Carlo method is adopted to analyze the effective emissivity of the radiant source. After a theoretical analysis on the distribution of the effective emissivity of the radiant surface, the source structure was constructed. Heat pipe technology keeps the source isothermal and the temperature control system ensures that the source is stable at a certain temperature. The radiant source has a broad applications in various fields, such as infrared imaging, infrared measurement and humidity test.

Although there are tens of ways to measure humidity, among which the most traditional methods are the dry and wet bulb thermometer whose precision is lower compared with modern electronic methods, most of these ways are not satisfactory in terms of precision and stability [7]. Capacitor sensors which possess fast response advantages are employed to test humidity, but their measuring precision is easily affected by electromagnetic interference. Besides, humidity testing is often affected by the environmental temperature, which is a factor that makes humidity sensors' precision not be high. Humidity testing through radiation possesses advantages of fast response and robustness with high precision. Our research on an infrared source which has highly sensitivity for humidity may provide an improved way to measure humidity.

2. Analysis on Characteristics of Radiant Surface

The Monte-Carlo method was utilized for analyzing the radiant surface with concentric V-grooves. Assuming that the surface is diffuse (Lambertian), the calculation on its luminance follows Lambert's cosine Law. The Monte-Carlo Model of thermal radiation was set up. A Monte-Carlo simulation is implemented through random sampling that is based on probability models whose deduction methods are based upon actual physical models. Exactitude methods like the Bedford method are just suitable to calculate simple symmetric cavity shapes. Although Monte-Carlo methods are flexible enough to be used for complicated cases, they are often regarded as unreliable methods with low precision. The effective emissivity of a cone was calculated by both the Monte Carlo method and the Bedford method, respectively. By comparing the results from the both methods, the correctness of the Monte Carlo method was proven.

2.1. Monte-Carlo Model in Thermal Radiation

The idea of a Monte-Carlo method is to set up a probability model or stochastic process whose parameters are equal to the solution of the problem to investigate, then to calculate the statistical characteristics of the required parameters through sampling, and the solution can be solved based on a vast number of observations. In thermal radiation calculations, local temperature and radiation fluxes are usually involved. The process of thermal radiation exchange is regarded as the movement of discrete energy beams. In this way, the local radiation flux can be obtained by calculating the number of beams that reach the local surface per unit time. These beams are deemed to consist of particles with a certain amount of energy. If the energy of each beam is equal, the local energy flux can be obtained by multiplying the number of beams arriving by the energy of each bundle of light per unit time per unit area. The direction of a transmitting beam i is determined by the direction angles θ, φ which are obtained by random sampling, as shown in Figure 1. dA is the area of a tiny piece. The radiation from dA has different intensity along with directions. For a diffuse surface, the emssivity of a surface is independent of the azimuth angle φ, but dependent on θ. The monochromatic emissivity of the surface is independent of the azimuth angle φ (The word monochromatic means the emissivity is under a certain wavelength λ, it is correspondent to the word 'total' that covers the whole spectrum, which means the wavelength λ is from 0~∞), and the total emission energy per unit time is as below:

$$E = \varepsilon\sigma T^4 dA \tag{1}$$

where ε is the material emissivity, T is the temperature of dA Kelvin, σ is Boltzmann's constant.

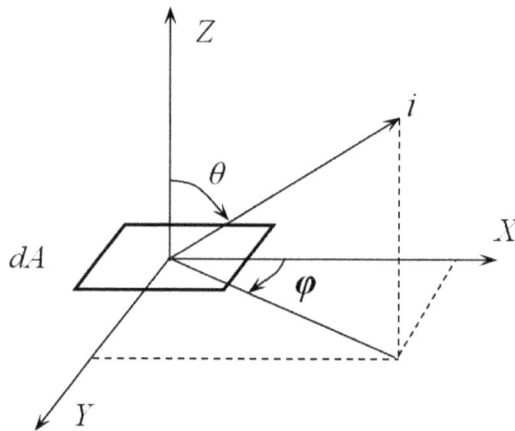

Figure 1. Direction of an emitted beam.

Assuming that there are two tiny pieces of blackbody dA_1, dA_2, and there exists radiation exchange between both the tiny pieces, then the angle coefficient F_{d1-d2} is defined by the ratio of the energy radiated out from dA_1 reached dA_2 to the total energy radiation of dA_1, as shown in Figure 2.

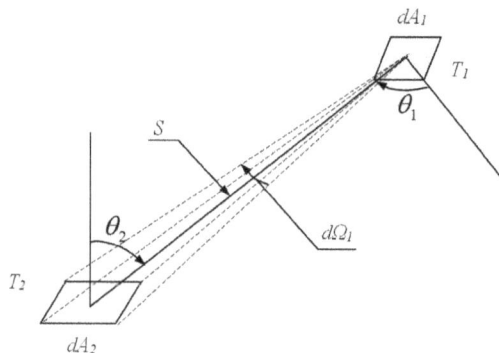

Figure 2. Radiant exchanging relation between two tiny surfaces.

When a Monte Carlo method is used in analyzing the characteristics of a blackbody cavity (as shown in Figure 3), the radiation of the light point i can be treated as two components, one is EF_i which radiates out of the aperture, the other is $E(1 - F_i)$ which goes to the wall at where it may be absorbed or reflected. If it is reflected then it is divided into two parts again which are $E(1 - F_i)F_{i1}$ and $E(1 - F_i)(1 - F_{i1})$, where F_{i1} denotes the angle factor between the point at where the beam from i is reflected first time. Assuming F_{ij} that denotes the angle factor between the point where the ray from i reflected at the time j, a beam of light can be traced in this way. When all the light points have been manipulated, the radiant flux out of the cavity and the effective emissivity are obtained [12]. However, for the V-groove concentric circles surface, the above model is not appropriate. The Monte-Carlo model is used to simulate the real physical model [13], in which radiant flux is coming from light points which are distributed uniformly along the surface. Each beam of emitting lights possesses the same amount of energy which is the total radiant energy of the point (it is an infinitesimal area dA), as shown in Equation (1). A light point's position is a that can be determined according to the geometry probability, and a ray's direction is determined by zenith θ and azimuthal φ. The event that a ray is reflected or absorbed is treated as a random variable [14]. If a surface is diffuse, the probability model of the random sampling θ is as follows:

$$P(\theta) = \int_0^\theta 2 \sin \theta \cos \theta d\theta = \sin^2 \theta \tag{2}$$

where r is random number. θ can be determined by generating r, φ is determined by $\varphi = 2\pi r$. All beams are traced thoroughly, and the evaluation of effective emissivity is totally based on the statistics of a tremendous number of samples, but simulated results are significantly affected by many factors such as the correctness of a Monte-Carlo model, the way light ray tracing is judged, sampling number, sampling method and the uniformity of the random number generator, etc.

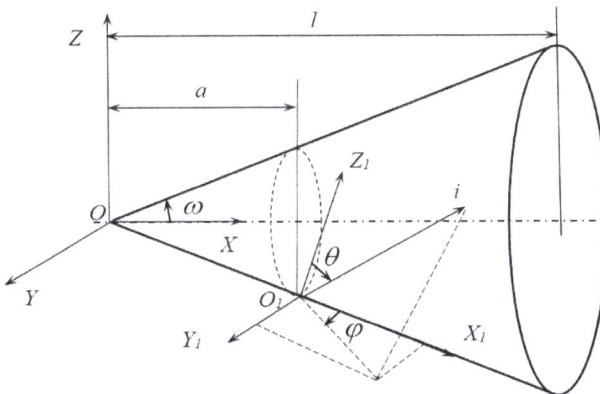

Figure 3. Light ray tracing in a cone cavity.

2.2. Monte-Carlo Simulation for a Cone Cavity

To calculate the effective emissivity of a cone cavity by the Monte Carlo method, the simulation procedure is as follows: the first step is to determine the position of a light point, a, (as shown in Figure 3) according to the area probability:

$$P(a) = \frac{\pi a^2 \sin \omega / \cos \omega^2}{A_{cone}} \tag{3}$$

where A_{cone} is the cone area, as a continue variable a, its random sampling is $r = P(a)$, where r is a random number, a is determined as follows:

$$a = \sqrt{\frac{r A_{cone} \cos^2 \omega}{\pi \sin \omega}} \quad 0 \le a \le l \tag{4}$$

The second step is to determine the beam's direction (θ, φ), θ random sampling is similar as a, according to Equation (2), the direct sampling method is as: $\theta = \sin^{-1} \sqrt{r}$, $\varphi = 2\pi r$.

In order to reduce the computing time, θ can be obtained by the rejection sampling method as follows: $\theta = r_1 \pi/2$ when $r_2 < \sin \pi r_1$. r_1, r_2 are random numbers.

The third step is to trace the beam, to judge where it goes, if it flies out of the cone, the accumulated radiant energy out E_{out}, if it is still in the cone, to judge that if it is absorbed or reflected, the cone equation in coordinate OXYZ is:

$$x^2 tg^2 \omega = y^2 + z^2 \tag{5}$$

The ray' equation in coordinates O1X1Y1Z1:

$$\begin{cases} z_1^2 = ctg^2\theta(x_1^2 + y_1^2) \\ \quad y_1 = x_1 tg\omega \end{cases} \tag{6}$$

The transformation between OXYZ and O1X1Y1Z1 is as shown below:

$$\begin{cases} x = x_1 \cos \omega + z_1 \sin \omega + a \\ z = -x_1 \sin \omega + z_1 \cos \omega - atg\omega \end{cases} \tag{7}$$

The crossing point of the ray and the cone is obtained from Equations (5) and (6), its coordinate in OXYZ is x (according to Equation (7)), if $x > 0$ and $x < 1$, the ray is still in the cone, otherwise it flies out, then the total energy E_{out} accumulates, then we go back to the first step.

If the ray is still in the cone, and when $r \le \varepsilon$, it is absorbed, and the program flow returns back to the original procedure to generate another new light point, otherwise $(r > \varepsilon)$ the ray is reflected. To take x as a, it goes back to the second step. When all the beams are traced, the effective emissivity $\varepsilon_a P(0)$ (0 represents the center cone) will be obtained as follows, A_{ap} is the area of the cone aperture:

$$\varepsilon_a P(0) = \frac{E_{out}}{\sigma T^4 A_{aP}} \tag{8}$$

2.3. Bedford Method for a Cone Cavity

The cone is divided into N segments along the axis. i represents the position and also the disk at the position i, di means the ring at i, F_{i-j} is the angle factor between disk i and disk j [15], it is the ratio of the radiant energy from disk i reaching disk j to the whole radiant energy of disk i, as shown in Figure 4, and it can be calculated in Equation (9):

$$F_{i-j} = \left\{ h^2 + r_1^2 + r_2^2 - \sqrt{(h^2 + r_1^2 + r_2^2)^2 - 4r_1^2 r_2^2} \right\} / 2r_1^2 \tag{9}$$

F_{di-dj} is the angle factor between ring di and ring dj, and F_{di-j} ring di and disk j. Other angle factors can be deduced, assume A_i, A_{di} are the areas of disk i and ring di, respectively, then $F_{i-dj} = F_{i-j} - F_{i-j+1}$, $F_{dj-i} = F_{i-dj} \times A_j/A_{di}$, $F_{dj-di} = F_{dj-i} - F_{dj-i+1}$, $F_{di-dj} = F_{dj-di} \times A_{dj}/A_{di}$, so the local effective emissivities along inside the wall of the cone, $\varepsilon_a(i)$ can be obtained as follows:

$$\varepsilon_a(i) = \varepsilon + (1 - \varepsilon) \sum_{j=1}^{N} \varepsilon_a(j) F_{di-dj} \tag{10}$$

To solve Equation (10), the initial values of $\varepsilon_a(i)$ in Equation (10) are set to ε. An iteration process is performed. When the errors $\varepsilon_a(i)$ between two adjacent iterations are less than 10^{-5}, the iteration process is stopped, and the convergence results $\varepsilon_a(i)$ are obtained. The results are as shown in Figure 5.

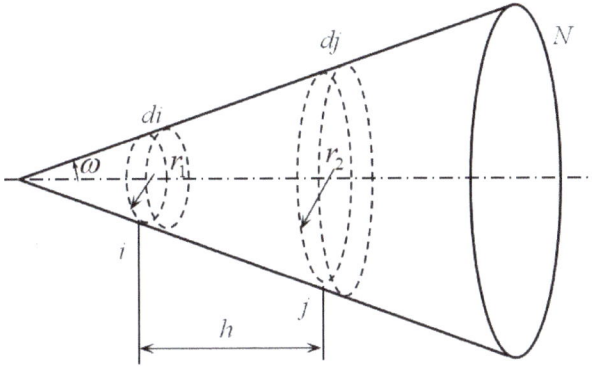

Figure 4. Radiation exchange in a cone cavity.

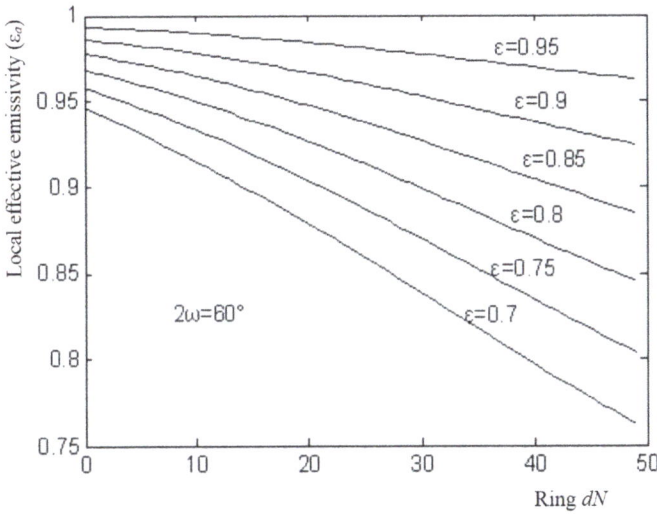

Figure 5. Local effective emissivity curves.

The effective emissivities of the cone $\varepsilon_a P(0)$ can be calculated from Equation (11). Curves corresponding to cone angles $2w$ reflect the changing of effective emissivities along with material emissivity ε which changes from 0.7–0.99 in 0.01 increments, and $2w$ from $10°\sim170°$ in $10°$ increments, as shown in Figure 6.

$$\varepsilon_a P(0) = \sum_{i=1}^{N} \varepsilon_a(i) A_{di} F_{di-N} \tag{11}$$

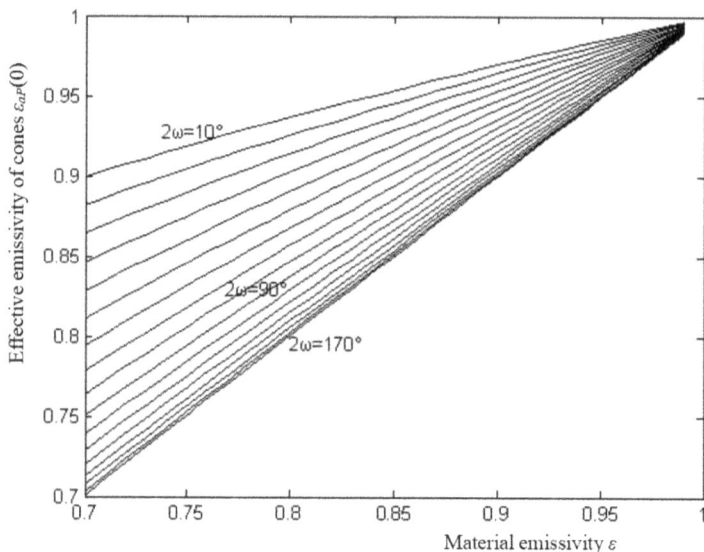

Figure 6. The effective emissivity of the cone.

2.4. Comparison of the Two Methods' Results

The exactitude methods such as the Bedford method are mostly used in blackbody radiant sources, but they are just suitable for sources with symmetric shapes such as a cone. As to the source with grooved surface, the Bedford method cannot be used, so Monte Carlo methods are developed. Monte Carlo methods are often deemed unstable and inaccurate, and their results fluctuate severely. The Monte Carlo method with large sampling number is expected to be more accurate. The effective emissivities of the cone $\varepsilon_a P(0)$ are calculated by both the Monte Carlo method and the Bedford method, respectively, along with material emissivity ε changing from 0.7~0.99 in 0.01 increments under the case when $2\omega = 45°$, and $60°$.

As an exactitude method, the calculation errors of the Bedford method are very small, within 10^{-5}, but the Monte Carlo method is totally based on random testing, and although it can even be used for complex calculations, it is often regarded as a method with poor calculation precision. The Monte Carlo computing results show converging results require large sampling number (N_s). The standard deviation is inversely proportional to $\sqrt{N_s}$. The results fluctuate severely when N_s is less then 10^{-6}, and they become stable when N_s reaches 10^{-7}. The Monte Carlo method needs very large samples, and for each calculation case, it needs 10^7 samples (number of light points N_s), and even up to 7×10^7. Comparing the computation of the two methods, the N_s of the Monte Carlo simulation is 2×10^7. The results show that the maximum calculation error between the Monte Carlo and the Bedford method is limited to 0.0004, and the errors for most calculation points (the discrete dots represent the calculations by the Monte Carlo method) are usually around 0.0001~0.0002, that means the results from the both methods coincide (as shown in Figure 7). The Monte Carlo method is therefore also a credible method in thermal radiation analysis. By comparing the results both from the Monte Carlo method and the Bedford method, the accuracy of the Monte Carlo method is verified to be high enough, close to the Bedford method, so it is proper to use the Monte Carlo method to simulate the V grooved radiant surface.

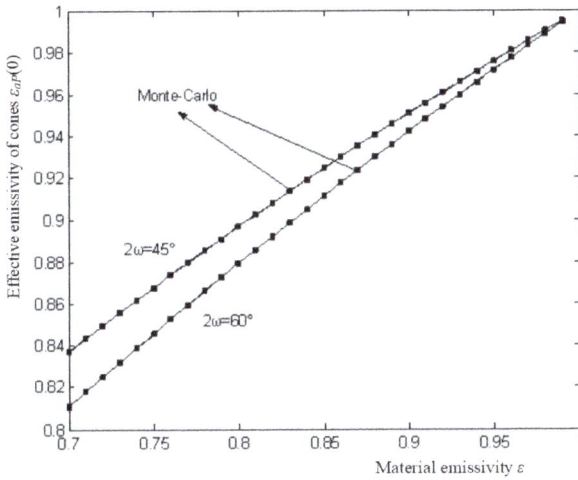

Figure 7. Comparison of the results between the two methods.

2.5. Monte Carlo Simulating for V-Grooved Surface

To increase the effective emissivity, the radiant surface is processed into concentric V-shaped grooves, as shown in Figure 8. Because it is too difficult to use the exactitude numerical method, the Monte-Carlo method is employed to simulate the radiation of the surface. The radiant energy is regarded to be composed of independent beams emitted by light points which are evenly distributed on the surface. These beams are completely traced through the Monte Carlo simulation.

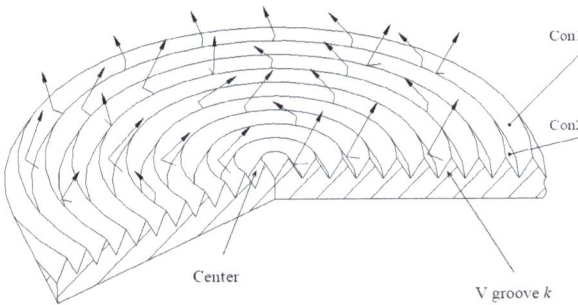

Figure 8. Surface with concentric V grooves.

To perform the Monte Carlo simulation, groove *k* is chosen, which is formed by two cones, the concave Con1 and the convex Con2, as shown in Figure 9. The first step is to determine the cone (Con1 or Con2) on which a light point is. When $r \leq A_{con1}/(A_{con1} + A_{con2})$, the light point is on Con1, otherwise is on Con2. A_{con1}, A_{con2} are the areas of the concave ring on Con1 and the convex ring on Con2 respectively. The second step is to determine the position of the light point. If it is on Con1, its position is *a*, if it is on Con2, its position is *b*. Both *a* and *b* are determined through random sampling as below:

$$a = L[k] - l + \frac{r(2L[k]-l)+3l-2L[k]}{2}$$
$$b = L_a[k] - l + \frac{r(2L_a[k]-l)+3l-2L_a[k]}{2} \tag{12}$$

where *r* is a random number. The next step is to determine the direction (θ, φ) of a beam, and to trace the beam [16]. The tracing process is similar to the case for the cone. The emission or reflection of each beam of lights obeys the probability distributions. The Monte Carlo simulation process is to trace each beam of light until it is absorbed or ejected out of the groove. The angle of emission of the beam, whether the beam is absorbed or reflected at the point of reflection and the angle of reflection are all regarded as random events and are determined by random sampling. When the number of light points is large enough, the statistical result of effective emissivity of the groove will converge to the true value. When the simulation of all V-grooves on the surface is completed, the effective emissivity distribution of the surface is obtained.

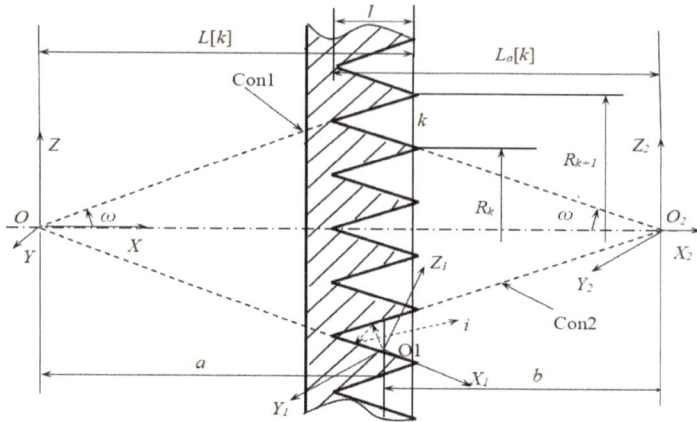

Figure 9. Light tracing for surface with concentric V-grooves.

The effective emissivity $\varepsilon_a P(k)$ of groove k is determined statistically after tracing all the rays. $\varepsilon_a P(k) = \varepsilon \cdot E_{out} \cdot (A_{con1} + A_{con2})/(A_r(k) \cdot N_s)$, where $A_r(k) = \pi(R_{k+1}^2 - R_k^2)$ is the aperture area of groove k, R_{k+1}, and R_k, represent radii of two adjacent circles which forms the aperture of groove k. The Monte-Carlo program was performed under the Visual Studio.net2003 environment. The distribution of effective emissivity $\varepsilon_a P(k)$ of a round piece surface was obtained when all the grooves had been computed ($\varepsilon = 0.95, \omega = 22.5°, l = 2.5$). When the number of sampled light points is less than 5×10^6, the results still fluctuate severely. The Monte Carlo calculation converges slowly [17], and when errors are less than 0.0002, the sampling number of light points is over 3×10^7. From the simulation results, the values of $\varepsilon_a P(k)$ become lower slightly from the center to the edge. Curve (1) represents the results corresponding to the sampling number $N_s = 3.5 \times 10^6$, and curve (2) is corresponding to $N_s = 3.5 \times 10^7$, as shown in Figure 10.

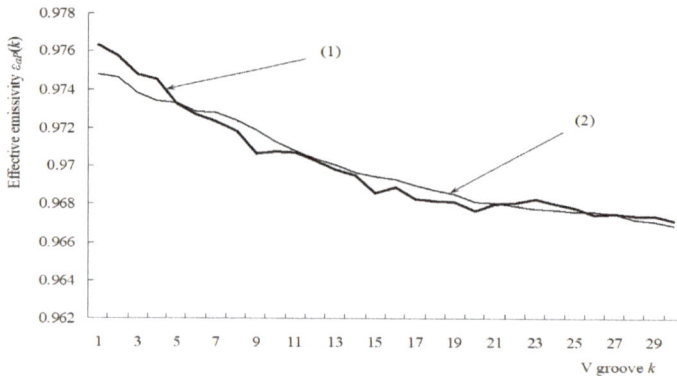

Figure 10. Distribution of effective emissivity of concentric V-groove surface.

The radiant source is composed of the V-grooved radiant surface, plus a cylinder to shield the circumference of the surface. The structure of the source is shown as in Figure 11. Heat pipe technology requires that the inside of the hollow shell is a capillary wicking porous material filled with heptane. This is an ideal area radiant source which can provide a uniform thermal radiation.

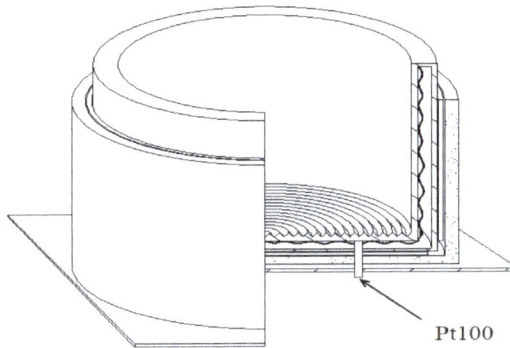

Figure 11. The structure of the radiant source.

3. Control System Design

The heating system possesses high heat inertia, and it is easily interfered by the environment. The fuzzy-PID control method is used in the temperature control system to make the system have a fast response, and operable across a large temperature range with high precision and strong robustness. The diagram of the control system is as shown in Figure 12. The primary task of the control system is to ensure the radiant source can reach a certain temperature in the temperature range of 5~95 °C. Heat pipe technology keeps the radiant source isothermal. Three Pt100 temperature sensors are installed on the heating surface to measure the temperature (as shown in Figure 11). A three-wire system is used to enhance the anti-interference ability. The temperature signals are amplified and transferred to a PIC16F876 microprocessor. The fuzzy-PID composite control method was adopted. The fuzzy control is appropriate for the large temperature range, and the PID is suitable to enhance the control precision [18]. The temperature signals are sampled by the A/D converter, and the control method decides the output. The power of the heater is controlled by a bidirectional silicon controllable rectifier (SCR) which is driven by the PWM outputs from the microprocessor. The incremental PID control is employed as shown in Equation (13) in which output values are decided by three adjacent sequential errors [19]:

$$\Delta u(k) = Ae(k) - Be(k-1) + Ce(k-2) \tag{13}$$

where A, B, C are coefficients. e means error. $e(k)$, $e(k-1)$, $e(k-1)$ are three sequential errors. All the control algorithms are based upon errors between the test value and the set value.

The fuzzy control method is widely used in temperature control systems [20], because it can imitate humans' judgment. E denotes the temperature error, which is the error between the set value and the measured temperature value. E_C represents the changes of temperature errors, and U denotes the fuzzy control outputs. Some symbols are used to represent the degree in a fuzzy set. NB denotes negative big, it expresses that deviation is negative big. NM negative medium, NS negative small, O zero, PS positive small, PM positive medium, and PB positive big. Assuming all the fuzzy variables' domains of E, E_C and U are {NB, NM, NS, O, PS, PM, PB}, the corresponding values are {−6, −5, −4, −3, −2, −1, 0, 1, 2, 3, 4, 5, 6}. Membership functions can be chosen as the triangle or Gaussian type, and the curves of the membership functions change along the fuzzy domain set. According to Zaden's 21 condition sentences, the fuzzy inferences are performed. If the error E is equal to NB or NM, and the change of errors E_C is NB or NM, then fuzzy output U should be PB under the fuzzy inference. The defuzzification of U can be got by Centroid method [18], as follows:

$$z_0 = \frac{\sum\limits_{i=1}^{n} \mu_{c'}(z_i) \cdot z_i}{\sum\limits_{i=1}^{n} \mu_{c'}(z_i)} \tag{14}$$

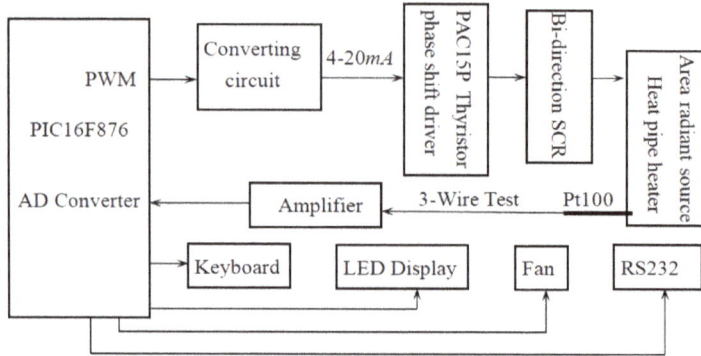

Figure 12. Control system with a PIC16F876 microprocessor.

The digital values of output U are obtained through defuzzification under all conditions of E, E_C. The output control table is formed by these values. Under a set of certain values of E, E_C, the output U can be determined directly by looking up the output control table whose values can be got by using Matlab [17], the procedures are as follows:

(1) Using the fuzzy control tool box of Matlab, we enter the FIS editor window, and in the Edit menu, select FIS Properties item, and because the temperature fuzzy control system is a two-dimensional system, E, E_C are chosen as input, and U is chosen as output.

(2) Editing the membership function of E, E_C and U, selecting the discourse domain of E to be $[-6, 6]$, selecting the membership function curves for the fuzzy subset {NB, NM, NS, ZO, PS, PM, PB} to be the Gaussian curve, and determining the gaussmf, parameter to be $[0.8493, -2]$, and for E_C and U the processes are similar.

(3) In the Edit menu, select the rules to edit, and the judging relationship between the two inputs E, E_C is set as "and". All fuzzy rule statements are entered. The fuzzy controller design is basically complete at this point, and the image corresponding to the input variables can be observed, and the process to determine the output variable value of the reasoning and the calculation process can be observed by the rule viewer, and through the surface viewer, the output surface map of U can be observed with the input E, E_C changes.

(4) The control file is saved as TemperContol, and the digital values of U are placed in the control output table [18].

The control output table $U[13][13]$ is the two-dimensional array whose capacity corresponds to the discourse domains of E, E_C. The range of a PWM register output values is 0~1023. When the fuzzy control is performed, the fuzzy values of E, E_C are used as the indexes of $U[13][13]$, and the values of discourse domains of E, E_C ($[-6, 6]$) should be converted into $[0, 12]$.

The temperature control flow is as follows: when the difference between the measured value and the set value is greater than the error limit (which can be determined by adjustments), the system is controlled by the fuzzy control, otherwise, it is under PID control. In the fuzzy control, when $E = -4$, $E_C = 3$, the output corresponds to $U[2][9]$ (as to E, -4 is converted into 2, E_C, 3 into 9), PWM output can be obtained by calling the function, set_pwm1_duty ($U[2][9]$).

4. Experiments

The radiation of this blackbody radiation source is experimentally verified by the optical system [21,22]. The radiation is confined to 8~12 μm far infrared light by the filter (see Figure 13), and then projected onto the HgCdTe infrared radiometer which performs a scanning test. Table 1 shows the results of the radiation field temperature under 40% RH laboratory humidity conditions. Figure 14 is a thermographic view of the scanning of the radiation source by the scanning radiometer. The standard deviation S_D of the measurement results corresponding to each setting temperature is less than 0.05 °C. With the 95% confidence interval, the uniformity of measurement results which can be obtained from the radiation source is within ±0.1 °C. Because the radiant surface of the source is isothermal, the uniformity of the testing value of the scanning radiometer verifies the uniformity of the effective emissivity of the radiant surface, which is consistent with the theoretical analysis.

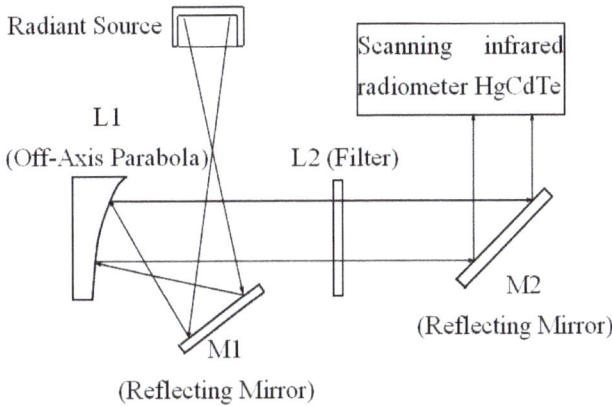

Figure 13. Optical test system for a radiant source.

The radiation source was scanned every 5 min at the temperatures of 40 °C and 70 °C, and the temperature field stability of the radiation source was obtained by statistical calculation of the measurement results.

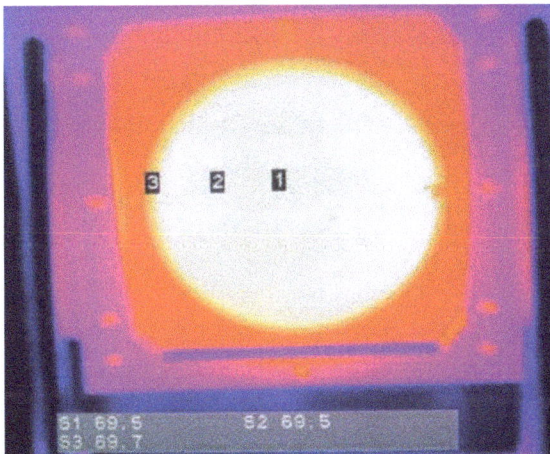

Figure 14. Thermal image of radiant source.

Table 1. Tested Results of the radiant source radiation under 40% RH.

T	$\overline{T}$ (°C)	ΔT_{max} (°C)	S_D (°C)
30.00	29.87	0.32	±0.03
40.00	39.00	0.45	±0.04
50.00	48.30	0.41	±0.05
60.00	57.98	0.31	±0.04
70.00	67.66	0.29	±0.04

The Planck blackbody radiation law (Planck's law) is the fundamental principle in infrared measurements. A blackbody's temperature can be calculated by simply measuring the blackbody's spectral emission power according to Planck's law [23]. In recent years, multi-spectral temperature testing technology which is based on Planck's law has been applied in practice. However, only radiometric parameters, such as radiance, are directly measured by the radiometer [24]. Through the corresponding test principle, the object radiation temperature can be deduced. In order to obtain the measured temperature, the emissivity of the measured object should be determined, but from another point of view, if the true temperature of the measured object is known, the performance of an infrared thermometer can be verified. The method of evaluating radiation source performance through radiation testing is presented here. If the temperature is determined by the two or more wavelengths of the specific temperature, the resulting color temperature of the object (T_s) is obtained. If it is based on the temperature of an object radiation temperature, a characteristic wavelength of radiation temperature, then the measured object temperature is the luminous temperature (T_l). An object radiation temperature, color temperature and luminous temperature are not the real surface temperature (T). When the real temperature of an object is T, its emissivity is $\varepsilon(T)$, the radiant output of the object is $M(T)$. When $M(T)$ is equal to the radiation of a blackbody whose temperature is T_τ, T_τ is called the object radiation temperature:

$$\varepsilon(T)\sigma T^4 = \sigma T_\tau^4 \tag{15}$$

when $\varepsilon(T)$ is known, the real temperature of the object can be derived from the radiation temperature T_τ according to Equation (15).

If the spectral emissivity of an object is $\varepsilon_\lambda(T)$, its spectral radiance is $L_\lambda(T)$. When $L_\lambda(T)$ is equal to the spectral intensity of a blackbody with temperature T_l, T_l is called the luminous temperature of the object. By the Planck's law, the calculation of the real temperature is simplified in Equation (16):

$$T = \frac{c_2 T_l}{\lambda T_l \ln \varepsilon_\lambda(T) + c_2} \tag{16}$$

where c_1 and c_2 are the coeffients in the Planck's law. $\varepsilon_\lambda(T)$ should be known in order to derive the real temperature of an object from the measured luminous temperature T_l. Similarly, if the spectral emissivities under the wavelengths λ_1 and λ_2 are $\varepsilon_{\lambda_1}(T)$, $\varepsilon_{\lambda_2}(T)$, respectively, the real temperature T can be determined according to Equation (17):

$$\frac{1}{T} - \frac{1}{T_s} = \frac{\ln[\varepsilon_{\lambda_1}(T)/\varepsilon_{\lambda_2}(T)]}{c_2(1/\lambda_1 + 1/\lambda_2)} \tag{17}$$

where T_s is the color temperature of the object. As to the measuring case, an object's emissivity, the luminous temperature or the real temperature of the object can be obtained by comparing the standard blackbody's temperature to the object radiation temperature [25]. If the emissivity of an object is known, its real temperature can be determined through comparing its radiation with a blackbody radiation. Conversely, if the real temperature of an object is known, its emissivity can be obtained [26]. Because the radiant source works in the wavelength band 8~12 μm, the following derivation is corresponding to a wavelength band. When the radiant source temperature is T and its spectral emissivity is $\varepsilon(\lambda)$, the spectral luminance on a radiometer is as follows:

$$E_T(\lambda) = \frac{1}{4}\varepsilon(\lambda)\tau_a(\lambda)\tau_0(\lambda)M_0(\lambda, T)\left(\frac{D}{f'}\right)^2 \tag{18}$$

where D and f' are the optical aperture and focal length of the optical system, τ_a, τ_0 are the spectral transmittances of the atmosphere and the optical system respectively, $M_0(\lambda, T)$ is a blackbody radiant flux. In the wavelength band $\lambda_1 \sim \lambda_2$, the detection system output signal level is as below:

$$U(T) = \frac{1}{4}A\left(\frac{D}{f'}\right)^2 \int_{\lambda_1}^{\lambda_2} R_V(\lambda)\varepsilon(\lambda)\tau_a(\lambda)\tau_0(\lambda)M_0(\lambda, T)d\lambda \tag{19}$$

where, $R_V(\lambda)$ is the detector's spectral response, and A is the area of the detector. When the test distance is fixed, the infrared transmittance is negligible. Assuming the average transmittance of the optical system in the infrared band $\lambda_1 \sim \lambda_2$ to be $\overline{\tau}$, $R_V(\lambda) = 1$ for an ideal detector, and the radiant source emissivity ε is independent of the wavelength, and the measured output level of the radiation source can be simplified as follows:

$$U(T) = \frac{1}{4}A\left(\frac{D}{f'}\right)^2 \varepsilon\overline{\tau}\int_{\lambda_1}^{\lambda_2} M_0(\lambda, T)d\lambda \tag{20}$$

Under the same test conditions, the output of the standard blackbody can be obtained. When the output level of the radiation source is equal to the standard blackbody output level, the emissivity of the radiation source is determined by the following equation:

$$\varepsilon = \frac{\int_{\lambda_1}^{\lambda_2} M_0(\lambda, T_b)d\lambda}{\int_{\lambda_1}^{\lambda_2} M_0(\lambda, T)d\lambda} \tag{21}$$

where, T_b can be used as the measured radiation source wavelength band, by comparing with the standard radiation source to determine. T is the real temperature, measured by the temperature measurement circuit. The radiant flux of the blackbody source can be determined by Planck's law, as follows:

$$M_0(\lambda, T) = \frac{\pi c_1}{\lambda^5[\exp(c_2/\lambda T) - 1]} \tag{22}$$

Substituting Equation (22) into Equation (21), the emissivity ε of the radiation source at temperature T can be determined. In order to avoid the complexity of deriving the integral solution in Equation (22), the emissivity ε is calculated by programming. The integral domain is from λ_1 to λ_2 ($\lambda_1 = 8$ µm, $\lambda_2 = 12$ µm, and the integration step is $d\lambda = 0.04$ µm). Table 2 shows the test and calculation results.

Table 2. Tested results of the radiant source emissivity.

T_b	29.805	39.165	48.385	58.03	67.68
T	30.03	39.94	50.05	60.02	69.96
ε	0.9964	0.9884	0.9765	0.9735	0.9713

The characteristics of a blackbody radiant source are often evaluated through theoretical analysis [27]. Theoretical evaluation on the effective emissivity has significance in the design of a radiant source. Material intrinsic emissivity ε is an important parameter in the evaluation of the emissivity of a radiant source, but it varies with the environmental conditions. This part is about the experimental research on the effective emissivity of the radiant source. The results of experimental tests show differences with the theoretical analysis of the effective emissivity. This is because that the theoretical analysis is based on the diffusion model in the whole spectrum band, and the practical measurements involve specular reflection in the 8~12 μm wavelength. There are differences between the actual situations and the ideal theoretical assumptions such as the isothermal and diffuse model, the size of the radiant source. As to the material intrinsic emissivity ε of the radiating surface, in the theoretical analysis, it is usually regarded as a constant, but in fact, ε itself also changes with the environmental temperature. If the effective emissivity of the radiation source calculated by theoretical analysis is adopted as the calibration parameter, it will inevitably produce large errors in the calibration of the actual infrared equipment, which will affect the infrared imaging quality [28]. From the analysis above, the uniformity of the emissivity of the radiation source should be verified by the uniformity of the radiation of the source in the testing spectrum band.

Fowler proposed the concept of black body mass (blackbody quality), and pointed out that the blackbody mass is dependent on wavelength and temperature, and it decreases with the increasing temperature [29,30]. The quality of blackbody is essentially a reflection of the radiation source, which is influenced by the radiant source structure, temperature, environment humidity, etc. This proves that humidity is sensitive to ultraviolet radiation [31]. The effective emissivity of a radiant source can be calibrated via a humidity sensor. After the calibration curve is obtained, it is easy to identify humidity by measuring the radiation of the radiant source. At a certain temperature, the emissivity of the radiation source should have some relationship with the changing environmental humidity. The calibration curve of relationship between the emissivity of the radiant source and humidity is obtained through calibration, which is implemented by the infrared radiant measuring process as the humidity changes at a temperature of 30 °C. Although the sensor DHT11 whose humidity testing precision is about ±5% RH suitable for 20~90% RH is not suitable to be used as for standard humidity measurements, here it is used just to show the radiance calibration method for humidity testing. The calibration curve for the emissivity of a radiant source changing with humidity is shown in Figure 15. The radiation measurement scheme is as shown in Figure 13, the data scanned by the radiometer were tested as the humidity changed and was measured by the DHT11. It can be found that the emissivity of the radiation source decreases rapidly with increasing humidity in the 30~50% RH range, and becomes smoothly downwards in the 50~70% RH range. Humidity can be measured by the radiation test on the radiant source, which possesses fast response and strong robustness compared with other methods. When the radiant source emissivity test value is 0.985, the humidity will be 43% RH with a test accuracy of ±5% RH, and ±1% RH resolution. The test precision is dependent on the calibration accuracy which is affected by the DHT11 humidity sensor calibration. If a highly precise humidity sensor such as the AM2301 were adopted, the testing accuracy would be improved. The advantage of the humidity test via the radiant source is its fast response. The humicap humidity sensor response time is usually more than 20 s, but the radiant humidity test takes 30 ms which is just the radiometer scanning time.

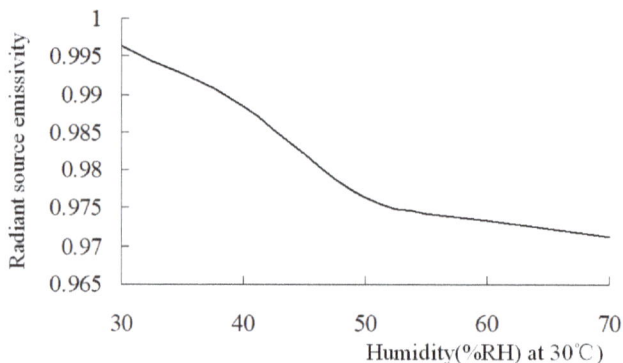

Figure 15. Calibration curve of the radiant source.

5. Conclusions

The characteristics of a surface radiant source were analyzed by the Monte Carlo method which is a flexible method that can simulate any real problem no matter how complex it is, but only if can it be described probably. The correct Monte Carlo model which has guaranteed simulation correctness should be established. Its convergence rate is expected to be very slow. To obtain accurate results, a large number of samples should be sampled. From the Monte Carlo simulation results, the samples should be more than 2×10^7. In addition, the random uniformity number and the correctness of the sampling method are also the important to obtain correct results. Besides the uniformity of random number and the reasonable model, the computation is also largely affected by the sampling number. The temperature control system of the radiant source has high thermal inertia and is susceptible to environmental temperature interference. The fuzzy-PID control method is incorporated with the heat pipe technology, which makes the system temperature be controlled at a high precision stable level with a rapid response. The characteristics of the radiant source were not only calculated theoretically, but also they were tested by a radiometer. The radiant spectrum band of the source was limited in the bandwidth of 8~12 μm via a lens filter. The results show that the temperature uniformity is within ±0.10 °C and the stability is within ±0.10 °C. A radiant source radiating out uniform energy can be used as a standard radiation source [18]. The radiant source can provide uniform thermal radiation over the radiant surface and radiate a stable radiant flux at a temperature in the range 5~95 °C, which can satisfy the demands for the calibration. Besides, the method of the calibration of the radiation of the radiant source is suitable to measure humidity, which is implemented by comparing the values of radiant flux which are obtained under both conditions of the laboratory humidity (e.g., 40% RH) and the real applied condition RH at the same temperature. Humidity measurements are easily affected by environmental factors such as temperature. Because humidity affects the radiation received by the radiometer, humidity can be determined indirectly by its radiation testing. Before performing a humidity test, a calibration is needed to obtain the calibration curve. The method of using a radiant source focuses on radiation measurement, which is just affected by humidity. The humidity testing accuracy of the method is largely dependent on the precision of the calibrating humidity sensor. Humidity sensors with humicaps often possess long response times, but the method which measures the radiation directly has a fast response. The method possesses advantages such as not being easily interfered by the environmental temperature, and a fast response. It is suitable for use in environmental humidity measurements in grain depots, greenhouses, warehouses, etc. The radiant source with a large radiant surface in the far infrared radiation region is therefore suitable for performing humidity tests at a near room temperature.

Author Contributions: H.Z. supervised the research. C.W. analyzed the data and wrote the paper. X.L. contributed to the literature review and helped to perform data analysis. B.S. analyzed the experiments and compiled the program. D.J. reviewed and edited the manuscript. All authors read and approved the final manuscript.

Funding: This research received no external funding.

Acknowledgments: This work is supported by Project of Natural Science Foundation of China (NSFC 51377037), Heilongjiang province innovative Project for undergraduate students (No. 201710214011).

Conflicts of Interest: The authors declare no conflict of interest.

References

1. Rubinger, C.P.; Calado, H.D.; Rubinger, R.M.; Oliveira, H.; Donnici, C.L. Characterization of a Sulfonated Polycarbonate Resistive Humidity Sensor. *Sensors* **2013**, *13*, 2023–2032. [CrossRef] [PubMed]
2. Boudaden, J.; Steinmaßl, M.; Endres, H.E.; Drost, A.; Eisele, I.; Kutter, C.; Müller-Buschbaum, P. Polyimide-Based Capacitive Humidity Sensor. *Sensor* **2018**, *18*, 1516. [CrossRef] [PubMed]
3. Kim, J.H.; Hong, S.M.; Moon, B.M.; Kim, K. High-performance capacitive humidity sensor with novel electrode and polyimide layer based on MEMS technology. *Microsyst. Technol.* **2010**, *16*, 2017–2021.

4. Ashis, T.; Sumit, P. Design and Development for Capacitive Humidity Sensor Applications of Lead-Free Ca,Mg,Fe,Ti-Oxides-Based Electro-Ceramics with Improved Sensing Properties via Physisorption. *Sensors* **2016**, *16*, 1135. [CrossRef]
5. Hashim, A.; Agool, I.R.; Kadhim, K.J. Novel of (polymer blend-Fe$_3$O$_4$) magnetic nanocomposites: Preparation and characterization for thermal energy storage and release, gamma ray shielding, antibacterial activity and humidity sensors applications. *J. Mater. Sci. Mater. Electron.* **2018**, *29*, 10369–10394. [CrossRef]
6. Kozlov, M.G.; Kustikova, M.A. Study of gas humidity sensors based on the absorption of vacuum ultraviolet. *J. Opt. Technol.* **2005**, *72*, 10–14. [CrossRef]
7. Lindauer, M.; Steinbrecher, R. A simple new model for incoming solor radiation dependent only on screen humidity. *J. Appl. Meteorol. Clim.* **2017**, *56*, 1817–1825. [CrossRef]
8. Hill, K.D.; Woods, D.J. The NRC blackbody-based radiation thermometer calibration facility. In *Temperature: Its Measurement and Control in Science and Industry*; AIP: Melville, NY, USA, 2003; Volume 7, p. 669.
9. Tonooka, H.; Sakuma, F.; Kudoh, M.; Iwafune, K. ASTER/TIR onboard calibration status and user-based recalibration. *Proc. SPIE* **2004**, *5234*, 191–201.
10. Bedford, R.E.; Ma, C.K.; Chu, Z.; Sun, Y.; Chen, S. Emissivities of diffuse cavities. 4: Isothermal and non-isothermal cylindro-inner-cones. *Appl. Opt.* **1985**, *24*, 2971–2980. [CrossRef] [PubMed]
11. Prokhorov, A.V. Monte Carlo method in optical radiometry. *Metrologia* **1998**, *35*, 465–471. [CrossRef]
12. Zhang, H.; Chu, Z.; Li, B. Monte-Carlo solution of characteristics of non-isothermal blackbody. *Infrared Res.* **1987**, *6*, 457–462.
13. Chen, G.; Kang, J. Frequency spectrum customization and optimization by using Monte Carlo method for random space vector pulse width modulation strategy. *Int. J. Signal Process. Image Process. Pattern Recognit.* **2016**, *9*, 135–152. [CrossRef]
14. Zhang, H.; Dai, J. A novel radiant source for infrared calibration by using a grooved surface. *Chin. Opt. Lett.* **2006**, *4*, 306–309.
15. Zhang, H.; Liu, B. Evaluation on radiant characteristics of a honeycombed surface by Monte Carlo simulation. In Proceedings of the International Forum on Strategic Technology, Ulsan, Korea, 13–15 October 2010; pp. 172–176.
16. Wang, Q.; Zhang, H.; Zhang, W.; Xie, X. Radiant characteristics evaluation and structural parameter optimization design of V grooved surface blackbody. *J. Tianjin Univ.* **2014**, *46*, 463–468.
17. Asl, S.V.; Davazani, Z.; Staji, S. Planning flying robot navigation in a three-dimensional space by optimaztion combining Q-learning and Monte Carlo algorithms. *Int. J. Hybrid Inf. Technol.* **2015**, *8*, 297–306. [CrossRef]
18. Zhang, H.; Zhu, H. Design and experimental test of a novel surface blackbody with honeycombs. In Proceedings of the International Forum on Strategic Technology, Ulsan, Korea, 13–15 October 2010; pp. 361–364.
19. Rafael, O.; Saul, O.J.; Roberto, M. Simulation of a temperature fuzzy control into induction furnace. In Proceedings of the 13th International Conference on Power Electronics (CIEP), Guanajuato, Mexico, 20–23 June 2016; Volume 8, pp. 64–69.
20. Boldbaatar, E.-A.; Lin, C.-M. Self-Learning Fuzzy Sliding-Mode Control for a Water Bath Temperature Control System. *Int. J. Fuzzy Syst.* **2015**, *17*, 31–38. [CrossRef]
21. Xu, J.; Meng, B.; Zhai, W.; Zheng, X. Calibration of common temperature blackbody based on thermal-infrared standard radiometer. *Infrared Laser Eng.* **2014**, *43*, 716–721.
22. Han, S.; Luo, W.; Hu, W.; Wang, R.; Han, Q. Research on infrared emissive area uniformity measurement of extended area blackbody. *Acta Opt. Sin.* **2013**, *33*, 112–116.
23. Xiu, J.; Jin, W.; Wang, X. Three-point infrared radiometric calibration and correction method using U-shaped blackbody. *Infrared Laser Eng.* **2013**, *42*, 2313–2318.
24. Wang, H.; Liu, H.; Ying, C.; Wu, B.; Huo, M.; Jiang, B. Testing of infrared spectroradiometer using blackbody. In Proceedings of the 12th IEEE International Conference on Electronic Measurement & Instruments (ICEMI), Qingdao, China, 16–18 July 2015; Volume 3, pp. 1289–1292.
25. Fentabil, M.A.; Daneshfar, R.; Kitova, E.N.; Klassen, J.S. Blackbody infrared radiative dissociation of protonated oligosaccharides. *J. Am. Soc. Mass Spectrom.* **2011**, *22*, 2171–2178. [CrossRef] [PubMed]
26. Hanssen, L.M.; Mekhontsev, S.N.; Zeng, J.; Prokhorov, A.V. Evaluation of blackbody cavity emissivity in the infrared using total integrated scatter measurements. *Int. J. Thermophys.* **2008**, *29*, 352–369. [CrossRef]
27. Hao, X.; Sang, T.; Pan, B.; Zhou, H. Research on characteristic parameter of Ta-Zro$_2$ fiber blackbody cavity temperature sensor. *Int. J. Signal Process. Image Process. Pattern Recognit.* **2016**, *9*, 287–296.

28. Te, Y.; Jesesck, P.; Pépin, I.; Camy-Peyret, C. A method to retrieve blackbody temperature errors in the two points radiometric calibration. *Infrared Phys. Technol.* **2009**, *52*, 187–192. [CrossRef]
29. Fowlder, J.B. An oil-bath-based 293 K to 473 K blackbody source. *J. Res. Natl. Inst. Stand. Technol.* **1996**, *101*, 629–637. [CrossRef] [PubMed]
30. Zhan, L.; Zhuang, Y. Infrared and visible image fusion method based on three stages of discrete wavelet transform. *Int. J. Hybrid Inf. Technol.* **2016**, *9*, 407–418. [CrossRef]
31. Wei, L.; Chao, D. Enhanced Humidity Sensitivity with Silicon Nanopillar Array by UV Light. *Sensors* **2018**, *18*, 660. [CrossRef]

Article

In-Depth Investigation into the Transient Humidity Response at the Body-Seat Interface on Initial Contact Using a Dual Temperature and Humidity Sensor

Zhuofu Liu [1,*], Jianwei Li [1], Meimei Liu [2], Vincenzo Cascioli [3] and Peter W McCarthy [4]

[1] The higher educational key laboratory for Measuring & Control Technology and Instrumentations of Heilongjiang Province, Harbin University of Science and Technology, Harbin 150080, Heilongjiang, China; 13904502205@163.com

[2] Department of Obstetrics and Gynecology, the Second Affiliated Hospital of Harbin Medical University, Harbin 150081, Heilongjiang, China; mm7723@163.com

[3] Murdoch University Chiropractic Clinic, Murdoch University, Murdoch 6150, Australia; v.cascioli@murdoch.edu.au

[4] Faculty of Life Science and Education, University of South Wales, Treforest, Pontypridd CF37 1DL, UK; peter.mccarthy@southwales.ac.uk

* Correspondence: zhuofu_liu@hrbust.edu.cn; Tel.: +86-139-0451-2205

Received: 30 January 2019; Accepted: 21 March 2019; Published: 26 March 2019

Abstract: Relative humidity (RH) at the body-seat interface is considered an important factor in both sitting comfort and generation of health concerns such as skin lesions. Technical difficulties appear to have limited research aimed at the detailed and simultaneous exploration of RH and temperature changes at the body-seat interface; using RH sensors without the capability to record temperature where RH is recorded. To explore the causes of a spike in RH consistently produced on first contact between body and seat surface, we report data from the first use of dual temperature and RH (HTU21D) sensors in this interface. Following evaluation of sensor performance, the effect of local thermal changes on RH was investigated. The expected strong negative correlation between temperature and RH ($R^2 = -0.94$) supported the importance of considering both parameters when studying impact of sitting on skin health. The influence of sensor movement speed (higher velocity approach: 0.32 cm/s ± 0.01 cm/s; lower velocity approach: 0.17 cm/s ± 0.01 cm/s) into a static RH region associated with a higher local temperature were compared with data gathered by altering the rate of a person sitting. In all cases, the faster sitting down (or equivalent) generated larger RH outcomes: e.g., in human sitting 53.7% ± 3.3% RH (left mid-thigh), 56.4% ± 5.1% RH (right mid-thigh) and 53.2% ± 2.7% RH (Coccyx). Differences in size of RH change were seen across the measurement locations used to study the body-seat interface. The initial sitting contact induces a transient RH response (duration ≤ 40 s) that does not accurately reflect the microenvironment at the body-seat interface. It is likely that any movement during sitting would result in similar artefact formation. As a result, caution should be taken when investigating RH performance at any enclosed interface when the surfaces may have different temperatures and movement may occur.

Keywords: humidity; transient response; body-seat interface; thermal impact; sitting rate; dual temperature-humidity sensor

---

1. Introduction

Sedentary behaviour has increasingly become the norm in many societies due to the increasing reliance on technology [1]. According to previous reports [2–4], it has been estimated that the average adult sits for up to two thirds of their time awake. However, extensive sitting postures along with low energy expenditure (physiological inactivity) have been strongly associated with many health problems such as obesity, type 2 diabetes, musculoskeletal symptoms and cardiovascular disease [5].

Another group that can be affected by too high humidity at the seat surface-skin interface is wheelchair users. This group often suffers from various skin diseases due to prolonged mechanical loading and in some cases, their inability to regulate air circulation in order to maintain their skin's viability. A typical and severe problem is the formation of pressure ulcers (PUs) when the skin is subjected to a mixture of persistent pressure, high humidity, raised metabolism and reduced blood flow. The yearly expenditure on PU diagnosis and treatment imposes heavy burdens on the health care sector. Between 1990 and 2001, it was reported that 114,380 people died from PU-related illnesses in the United States [6,7] alone. Although the cause of PUs is multifactorial, prolonged moisture and mechanical loading have been considered the main factors underpinning their formation and development [8].

Regarding the relationship between PUs and moisture build-up at the skin-seat/skin-mattress contact surface, a mathematical model was established to optimise the microclimate factors including relative humidity (RH), temperature (T) and pressure (P) [9]. From the perspective of aetiology, the causation of PUs in chronically immobilised patients was investigated and the results indicated that ensuring the flow of water vapour from the patient's body to the outside was an essential component in the prevention of skin lesions [10,11]. RH has been identified as a primary cause of PU formation together with lack of activity, friction and shear [12]. As there are no obvious symptoms prior to the occurrence of PUs, monitoring the microclimate characteristics at body-seat interface may be an effective tool to help reduce its prevalence.

A number of developments have been reported in order to allow the testing of designs, including a sitting simulator to evaluate the microenvironment properties of wheelchair cushions [13] and a personalised seat ventilation system capable of compensating for the RH deficit in a relatively high-density space (aircraft cabins [14] or classrooms [15]). To evaluate the performance of moisture dissipation, the TRCLI (thermodynamic rigid cushion loading indenter) was developed and used to compare the capabilities of different wheelchair cushions in terms of dissipating water vapour [16]. Furthermore, wheelchair cushion selection criteria have been proposed taking the transfer of water vapour into consideration [17].

Although RH has been recognised as playing an important role in the formation and development of PUs, it has not proved easy to measure reliably without the researcher having to directly influence the environment and can be prone to be influenced by the adjacent environment. We have performed sitting experiments in a wide range of ambient conditions (damp summer days in the UK and very dry cold winter days in China) [18,19]. Although experimental room conditions were similar in terms of temperature, obvious differences existed in ambient moisture and, as a result, background RH. Regardless of these differences, the profile of RH change on sitting and remaining seated generally appeared very similar. Although much of the experiment produced expected results, one segment of the RH sensor output regularly caused us to pause for thought, namely the initial recording period. This period focuses on the duration directly related to sitting down and was associated with an apparent spike in RH [18]. In previous recordings, we were using humidity sensors without any direct temperature assessment. Although we had recognised the spike to most likely be an artefact related to our methodology, another researcher in the field (Siyu Lin) reported seeing a similar artefact. On raising the question of cause, we decided to explore the transient response of RH characteristics at the contact surface. The rationale for this was two-fold. Firstly, to understand the changes occurring at this point and attempt to describe any artefact components as they may mask other potentially important data; secondly, to ensure an understanding of why this occurred as it might also be capable of affecting recordings made later in any sitting session. Fortunately, the timing of this research allowed us to be first to report the use at the skin-seat interface of a sensor chip, which had both RH and temperature measurement capability. This chip gave us, for the first time, the opportunity to accurately co-locate these two measurements and, therefore, look to the degree of relationship between them.

Therefore, the purpose of this study was to offer comprehensive insight into the RH varying patterns over time following the user's initial contact with the seat.

2. Materials and Methods

2.1. Hardware System Description

The main component of the data acquisition unit is an ATmega328 processor (Microchip Technology, Chandler, AR, USA) which is connected to the HTU21D sensors (capable of simultaneously acquiring RH and temperature information: TE Connectivity Ltd., Rheinstrasse, Schaffhausen, Switzerland) through an IIC (Inter-Integrated Circuit) interface. Collected data are transferred from the processor to the computer through a USB cable and stored in the computer's hard drive disk for further off-line analysis. The sampling interval chosen is 10 Hz as the rate of temperature and RH change at the body-seat interface is much lower than this frequency [18,19].

2.2. Sensor Evaluation

Prior to conducting any experiment, a freshly acquired, commercially available HTU21D sensor randomly selected from a recently purchased package was assessed to determine its performance, particularly accuracy and linearity along with repeatability and hysteresis. A standardised environmental chamber (PVS-3KP, ESPEC Environment Equipment Co. Ltd., Hudsonville, MI, USA) was used for this purpose, having the capability of providing reliable RH variations from 10% RH to 95% RH (Certificate No: ISO 04308Q11746R0 M and EN AC/0708030).

The RH range of the chamber was manually changed to 13–93% RH, with increment/decrement steps of 4% RH. The chamber temperature was set to 25 °C $\pm$ 0.1 °C, which is the recommended testing temperature for the sensor in accordance with the manufacturer's data sheet. At each testing point, the stabilised sensor's output was recorded and comparative tests were carried out using the averaged value of five outputs (at each stabilised testing point) from the sensor. In addition, RH increment/decrement tests were performed five times in order to examine the repeatability.

2.3. Consistency Test

Three HTU21D sensors were attached to the contact surface of the foam cushion in the approximate locations suggested by previous research [18,19]. Briefly, these are: one on each side of the cushion symmetric to the central line (approximating the middle of each thigh) and a further sensor at the rear (at a point approximating the location of the coccyx region). To replicate the influence of pressure imposed on sensors due to human sitting, dummy buttocks were created. They were made by inserting sand bags into a pair of jeans with the lower legs sewn shut. The total mass of the calibration dummy buttocks was 50 kg in order to mimic the human loading introduced to cushions used in previous studies [19–21].

Attention was paid to ensure that all the sensors would be fully covered by the dummy buttocks in the loading trial. A 1-h duration trial was conducted in a vacant research room with environmental conditions: 27.8 °C $\pm$ 0.1 °C and 39.9% $\pm$ 0.2% RH. The door of the research room was closed during the entire experiment with no researchers remaining in the room during this time. This precaution was to limit the likelihood of any possible disturbance to the room temperature and RH levels.

2.4. Heating Trial

To examine the influence of thermal variations on the sensors RH output, an experiment was carried out to imitate body-seat interface RH changes at the initial point of sitting down under the environmental condition: 43.0% $\pm$ 1.4% RH and 22.5 °C $\pm$ 0.2 °C. An HTU21D sensor was placed in a cavity (63.3 $\times$ 40.4 $\times$ 18.7 mm: Length, Width and Depth, respectively) cut into the surface of a foam seat pad (cushion). An aluminium water tank (223 $\times$ 132 $\times$ 66 mm: Length, Width and Depth, respectively) containing half its total volume of tap water was then positioned on the top of this slot.

Water temperature was monitored by reading from an Hg thermometer in the water tank. Boiled water was gradually added to the water already in the tank until the water temperature in the tank reached 40 °C. The water was then allowed to cool until the temperature in the water tank dropped to around 20 °C. During the whole process, both temperature and RH values inside the slot in the foam cushion were recorded by the HTU21D sensor and transmitted from the digital communication interface of the microprocessor to the computer through a USB cable.

2.5. Human Trial

A healthy university student (174 cm and 58 kg) voluntarily participated in all tests, which had been approved by the ethics committee affiliated to Harbin University of Science and Technology. The participant gave written informed consent prior to volunteering. To prevent any effect caused by the material of any garment worn, the participant was asked to wear jeans while taking part in all trials. Measurements were made in the same laboratory as the other tests with both ambient temperature and RH being continuously monitored. The laboratory door was closed during the testing period to avoid any disturbance to the environmental conditions.

2.5.1. Sensor Movement

Presuming that differences in approach speeds between body and seat interface could have an impact on the transient RH variance, an adjustable speed system (Figure 1) was developed on which the sensor was mounted. During the trials, the body was kept in contact with the seat surface (relative static) while the sensor approached the contact surface at different speeds. The sensor was not fixed in the hole for the following reasons: (1) approach velocities between body and sensor would not be consistent between each trial; (2) the sensor's movement was considered equivalent to the movement of the participant according to relative movement theory [22]. Additionally, it was possible to adjust the moving speeds precisely by programming the microprocessor.

Based on the above design considerations, the adjustable speed system consists of a stepping motor, a rack and pinion, a control unit based on ATmega328 processor (Microchip Technology, Chandler, AR, USA) and some peripheral circuitry. The HTU21D sensor was fastened to a custom-made rack and pinion platform using hot melt adhesive (Delixi Electric Ltd., Zhejiang, China). A hole (20 mm in diameter) was created in a foam cushion fixed on a commercially available wooden chair. The hole was made at approximately the position that the middle part of the left thigh would be expected to occupy in a seated person. The hole pierced the whole of the foam and seat below. The sensor was driven upwards through the hole from below the chair towards the uppermost surface of the foam.

Testing was divided into two trials according to the sensor's movement speed (0.32 cm/s ± 0.01 cm/s and 0.17 cm/s ± 0.01 cm/s). Each trial was repeated 10 times, with a 20-min recovery interval being provided between each trial in order to ensure the sensor had similar starting conditions. As previously, in order to reduce the impact of environmental changes, the whole test was completed in the same laboratory, with the door closed throughout the recording period. The ambient temperature and RH of the laboratory during these trials were 28.5 °C ± 0.2 °C and 57.9% ± 0.9% RH, respectively.

After initialising the system, the participant was asked to take a seat. The sensor system was then programmed to move up through the hole at the preset speed. When the sensor came into contact with the left mid-thigh (measurement position), the stepping motor stopped. During the whole process, RH values were recorded and the last five recordings were averaged to represent the immediate contact RH value.

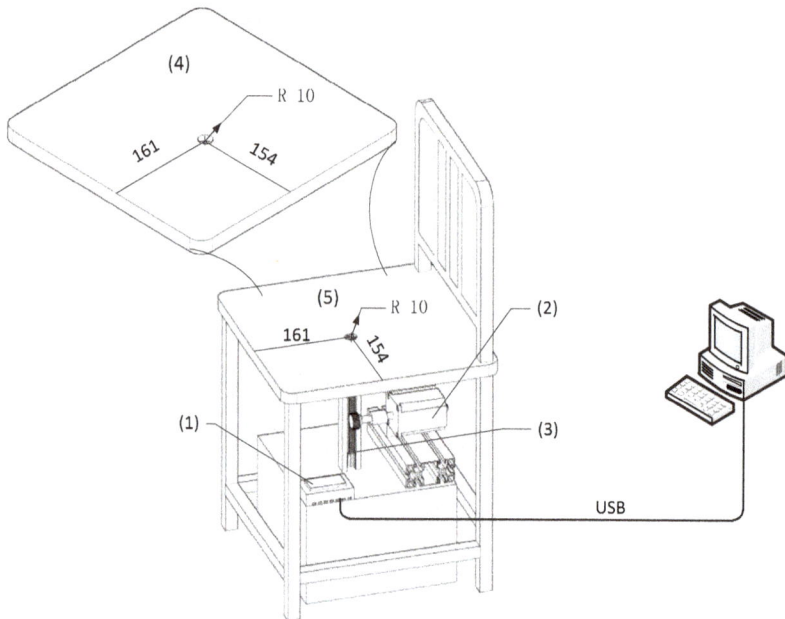

Figure 1. Configuration of the sensor movement system. Major components included: (1) data acquisition unit, (2) stepping motor, (3) transportation mechanism, (4) custom-made foam cushion with a drilled hole and (5) the wooden chair surface configured with the same hole. Both the customised cushion and the chair had equal dimensions (399 × 378 mm, length and width, respectively).

2.5.2. Participant Movement

In this part of the trial, three HTU21D sensors were fixed to the foam cushion at three sensitive locations described above: left mid-thigh, right mid-thigh and coccyx relative to the body structure of the participant [18]. To evaluate the effect of sitting speed on the interface RH, the same participant sat down at two different speeds (slowly or rapidly). Firstly, the participant stood in front of the test chair and waited for the order to sit down. The data acquisition system was activated to record the initial RH values, which were used as reference baselines. At this point, the participant was asked to "take a seat" (either slowly or quickly). After the participant sat properly (the buttocks were fully in contact with the foam cushion), the data acquisition process was manually terminated. The sitting speed trials were repeated ten times with a 20-min time gap between trials to allow sensors to recover to a condition similar to that at the start. The ambient temperature and RH parameters during these experiments were 24.7 °C ± 0.2 °C and 38.6% ± 0.4% RH, respectively.

2.6. Statistical Analysis

Normality of the data was assessed using Kolmogorov-Smirnov tests followed by appropriate paired test (t-test) to investigate the influence of sitting speed on RH at the sensor site. One-way analysis of variance (ANOVA) was used to examine RH characteristics among different measurement locations (participant movement trials), followed by a post-hoc Tukey-Kramer test [23] where appropriate. Along with the ANOVA, a Tukey-Kramer test is effective at determining the difference among each series of reading by comparing all possible pairs of means. The significance level was set as $p < 0.05$ for all statistical analyses. Finally, correlation analyses were performed to determine potential relationships between obtained measures.

3. Results

3.1. Sensor Performance

Figure 2 illustrates the sensor's performance in terms of linearity, repeatability and accuracy using data from the environmental chamber trials where RH was increased and then decreased, at the same temperature. Regarding accuracy, the absolute maximum difference between average (n = 5) reported value and the standardised RH values (environmental chamber output) was 1.8% RH over the full testing range. Regarding repeatability, the absolute maximum deviation among the five tests was 0.5% RH.

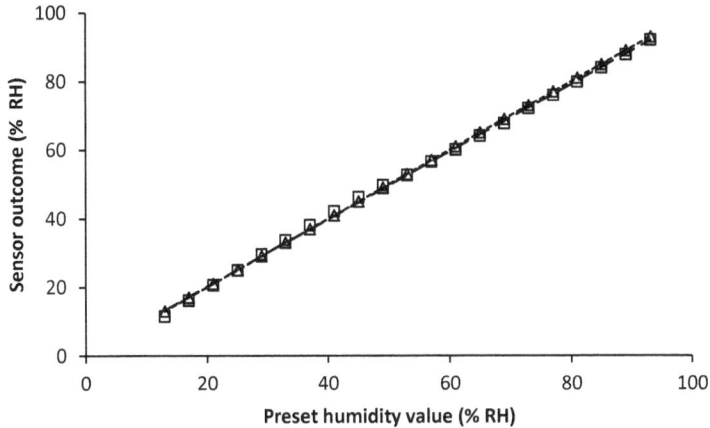

Figure 2. Performance of the humidity component (RH) of the HTU21D sensor. The square shape (□) indicates the measurements associated with increments in RH while the triangle shape (△) indicates measurements associated with decrement changes. The dashed line is the fitting curve for the increment data and the dotted line is the fitting curve based on the decrement data. The fitting equation for the increment test is y = 0.984x + 0.5124 while for the decrement test it is y = 1.0111x + 0.4142.

Output from the sensors also exhibited an approximately linear relationship in relation to that from the standardised environmental chamber (R^2 = 0.9989 and 0.9987 for increasing and decreasing RH trials, respectively). In terms of hysteresis resulting from increasing and decreasing RH, the correlation between RH outputs from the sensors was R^2 = 0.9997.

As three sensors were to be used in trials, their outputs were compared (Figure 3) under a sitting simulator (to simulate static loading pressure of sitting) and in a constant humidity/temperature environment for 1 hour. Outputs from the three sensors (Mean ± 1SD) were 41.4% ± 0.2% RH, 41.0% ± 0.2% RH and 41.1% ± 0.2% RH, respectively. A one-way ANOVA revealed no significant difference between the three sensors (p > 0.1).

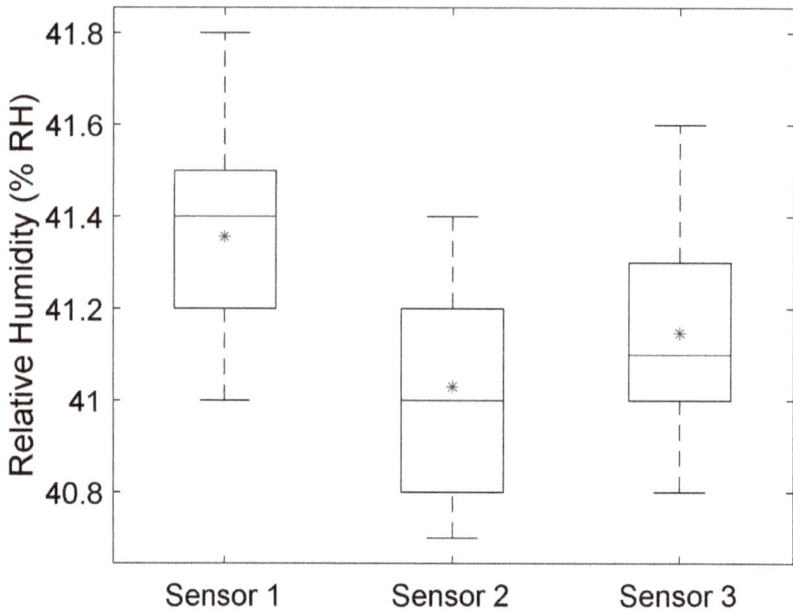

Figure 3. Box and Whisker plot of the data from the consistency test for the three humidity sensors using dummy buttocks to imitate the body pressure on the cushion surface. The experiment was performed in a vacant laboratory over a period of one hour (environmental temperature and RH: 27.8 °C ± 0.1 °C and 39.9% ± 0.2% RH). Top and bottom whiskers on the figure represent the maximum and minimum values for the corresponding humidity sensors, while the line inside each box represents the median value. The upper and lower borders of the boxes represent the 75th and 25th percentile values, respectively. In addition, the average values have been indicated by '*' in the boxplot.

3.2. Relationship between Temperature and RH

The influence of temperature changes on RH within a small region was assessed under the following environmental conditions: 43.0% ± 1.4% RH and 22.5 °C ± 0.2 °C (Figure 4). During the heating stage (hot water filled into the tank causing the air temperature inside the slot of the foam cushion to steadily rise due to thermal exchange), RH values within the region dropped to the lowest points (17.2% RH). In the natural cooling stage (the water was left to cool down without any interference), RH values gradually increased and stabilised at 45.2% RH. The correlation coefficient between temperature and RH value was −0.94.

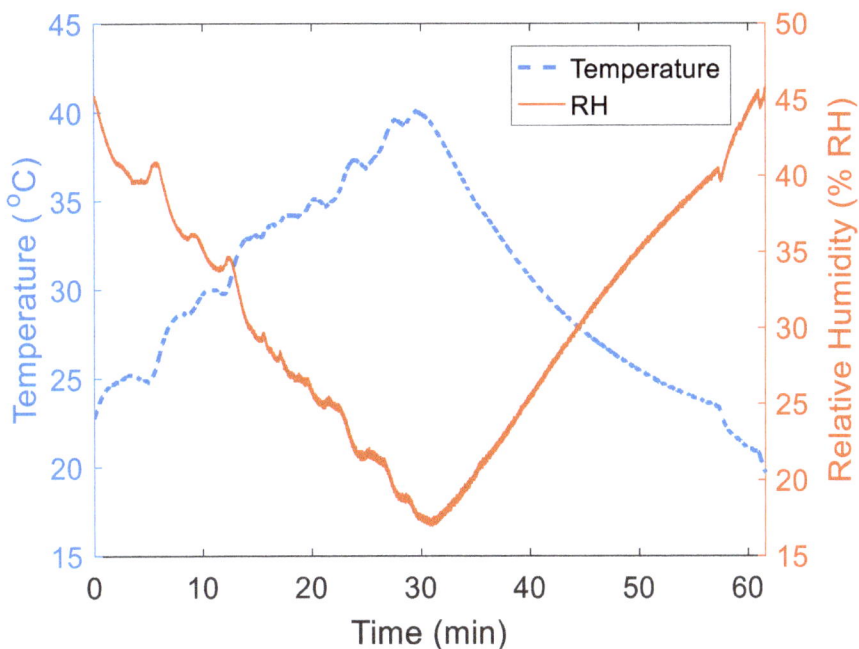

Figure 4. Graphical representation of the one-hour heating trial to illustrate the relationship between temperature change and relative humidity (RH) inside the relatively sealed environment created within the slot in the foam cushion. The dashed line represents the variation in temperature values (created by changing water temperature inside a metal container, which was monitored by reading from an Hg thermometer) and the solid line is the response of RH within the enclosed space in the foam cushion slot (environmental condition: 43.0% ± 1.4% RH and 22.5 °C ± 0.2 °C).

To further analyse the relationship between the temperature and RH (Figure 5), Heat Index (HI) was calculated using the following equation [24]:

$$HI = c_1 + c_2 T + c_3 R + c_4 TR + c_5 T^2 + c_6 R^2 + c_7 T^2 R + c_8 TR^2 + c_9 T^2 R^2 \tag{1}$$

where:

$c_1 = 0.363445176,$	$c_2 = 0.988622465,$	$c_3 = 4.777144035$
$c_4 = -0.114037667,$	$c_5 = -8.50208 \times 10^{-4},$	$c_6 = -2.0716198 \times 10^{-2}$
$c_7 = 6.87678 \times 10^{-4},$	$c_8 = 2.74954 \times 10^{-4},$	$c_9 = 0$

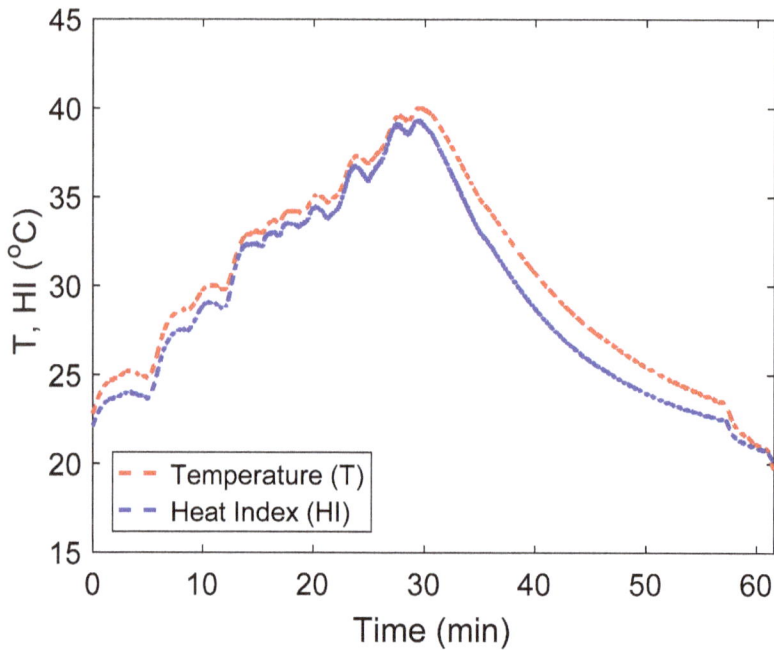

Figure 5. Comparison of heat index (HI) and measured temperature (T), both reported as °C, where the red line represents the measured temperature values and the blue line is HI calculated using relative humidity (RH) and the temperature from within the slot in the foam cushion.

To examine how RH varies according to the changing temperature, Dew Point (DP) values were also calculated [25]:

$$DP = \frac{243.04 \times \left(\ln\left(\frac{RH}{100}\right) + \frac{17.625T}{243.04+T} \right)}{17.625 - \ln\left(\frac{RH}{100}\right) - \frac{17.625T}{243.04+T}} \tag{2}$$

The calculated DP values were compared to a DP table (https://www.lamtec.com/technical-bulletins/dew-point-table/) utilising three testing points (starting point, near peak and end point) to examine RH variation in relation to changes in temperature.

3.3. Measurement of the Body-Seat Interface RH

3.3.1. Sensor Movement

In the sensor movement tests (n = 10), significantly different (p < 0.05) RH values at the body-seat interface were found between the fast (85.0% ± 1.6% RH) and slow (78.5% ± 3.5% RH) velocity of the stepping-motor-driven sensor system (Figure 6).

To analyse the relationship between the sensor approach speeds and RH, the average values of each trial for sensor speed and its relative RH were assessed for the presence of a correlation. However, no correlation was observed between high speed and RH, whereas at best a weak correlation might exist with slow speed (correlation coefficients 0.02 and 0.64, respectively). In addition, the trend curve for both moisture and temperature showed an apparent "spike" after the sensor achieved full contact with the left mid-thigh of the participant (Figure 7).

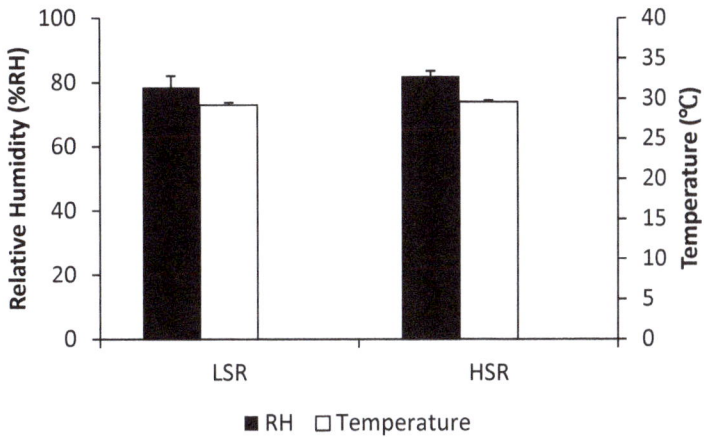

Figure 6. Comparison of body-seat interface relative humidity (RH) when the sensor moves towards the left mid-thigh of the participant at two different velocities: lower speed rate (LSR; 0.17 cm/s ± 0.01 cm/s) and higher speed rate (HSR; 0.32 cm/s ± 0.01 cm/s). In addition, the temperature values at the contact surface for different speeds of sensor movement are illustrated along with RH. Error bars denote ±1 standard deviation.

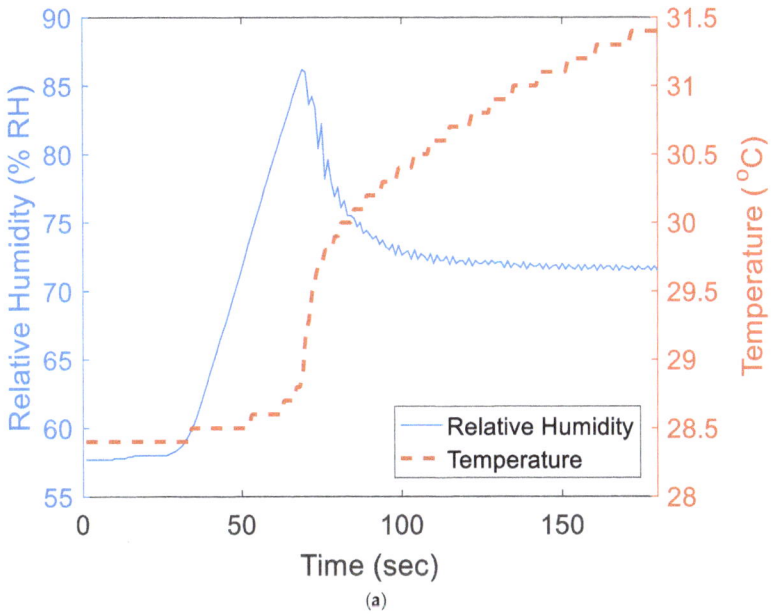

(a)

Figure 7. *Cont.*

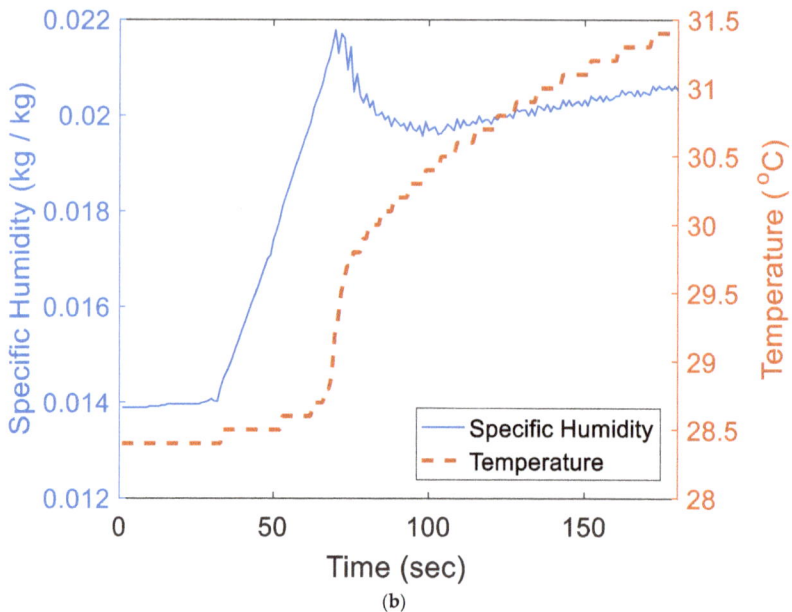

Figure 7. Varying patterns of temperature and moisture based on a data set from one of the 10 repeat trials. After the sensor was fully in contact with the mid left-thigh of the participant, the system remained in that position to record information for approximately 100 seconds: (**a**) relative humidity (RH) and temperature (T); (**b**) specific humidity (SH) and temperature where SH data were estimated using a humidity converter from the following website (http://www.humcal.com/index.php).

3.3.2. Participant Movement

When the participant sat down either slowly or rapidly, the body-seat interface RH responded differently at the measurement locations (Figure 8). Based on the 10 repeat trials, the statistical values (Mean $\pm$ 1SD) were 51.0% $\pm$ 1.6% RH (left mid-thigh), 51.8% $\pm$ 1.3% RH (right mid-thigh) and 48.5% $\pm$ 0.5% RH (Coccyx) for sitting slowly. In comparison, the fast sitting down generated larger RH outcomes: 53.7% $\pm$ 3.3% RH (left mid-thigh), 56.4% $\pm$ 5.1% RH (right mid-thigh) and 53.2% $\pm$ 2.7% RH (Coccyx).

Significant differences were found among different sitting speed for each measurement location ($p < 0.05$: paired t-tests). The sitting speed in the sagittal plane was estimated by measuring the vertical and horizontal distances between the bottom (initial contact point) just before sitting and the contact point on the seat cushion. The trajectory between these two points was estimated by assuming a right angled triangle and calculating the hypotenuse in order to approximate the moving path during sitting down process (linearity was assumed for ease of calculation, although it is acknowledged that the path was likely to be slightly curvilinear). The time taken to sit for each participant movement trial was recorded and divided by the estimated distance travelled.

Another interesting discovery was that significant differences also existed among the different measurement locations at the lower sitting speed ($p < 0.05$), while the difference was not significant for the faster sitting speed ($p = 0.13$). To further explore this apparent difference, a Tukey-Kramer test was applied to analyse the RH data from the slow sitting down experiments. Results indicated that the coccyx region sensor showed a significant difference from both mid-thigh measurements (the difference between left mid-thigh and coccyx was 2.4% $\pm$ 1.5% RH and it was 3.3% $\pm$ 1.5% RH between right mid-thigh and coccyx), while there was no significant difference between left and right mid-thighs.

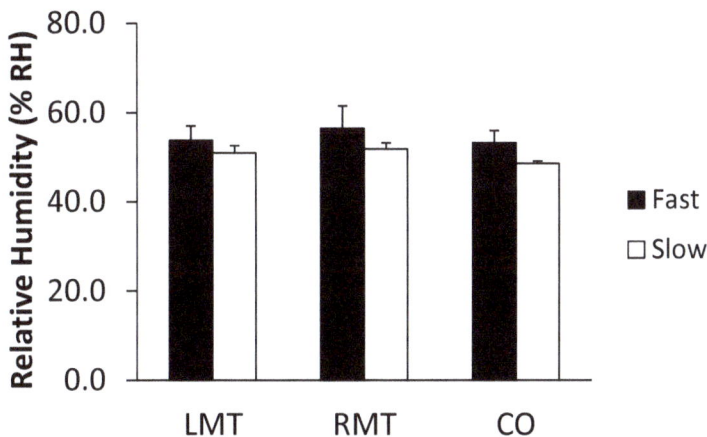

Figure 8. Averaged relative humidity (RH) values based on 10 sets of measurements including the last five measuring points from each of the three sensitive locations: left mid-thigh (LMT), right mid-thigh (RMT) and coccyx (CO), when the participant sat down at different speeds (fast: 10.29 ± 0.90 cm/s; slow: 6.78 ± 0.43 cm/s). Error bars denote the first standard deviation. The ambient temperature and RH parameters were 24.7 °C ± 0.2 °C and 38.6% ± 0.4% RH, respectively.

4. Discussion

4.1. Sensor Evaluation

Based on the sensor performance tests, it is conceivable that HTU21D is more suitable for detecting body-seat interface microenvironment parameters than traditionally used single modality sensors [18–20]. Firstly, the accuracy is ± 2% RH along with lower hysteresis (±1% RH) according to the sensor's datasheet, confirmed in part by the findings presented here. Secondly, it integrates both temperature and RH detectors in a single tiny microelectronic chip (3 × 3 mm). As a result, it is possible to measure both thermal and humidity information at the same contact area without the need of deploying two solo-functioning (temperature/humidity) transducers. The thermal detection range is from −40 °C to 100 °C with the capability for full range RH measurement (0% RH to 100% RH). Lastly, but not least, the price for a breakout board of HTU21D is minimal, making it financially and practically possible to construct a sensing array for use at the contact surface similarly to our previously published research [20].

4.2. Transient Characteristics of RH at the Contact Surface

Based on the experimental results presented here, we conclude that temperature variance can and does influence the recorded RH during the initial period when the surfaces move into contact A further factor appears to be the speed of the sensor approaching the buttocks and vice versa (i.e., speed of sitting down). RH is directly related to the temperature of the sensor when it makes the reading, therefore, knowledge of the temperature would be considered important. When two surfaces at the same RH but different temperatures come into contact, the sensors may misreport the water vapour before the temperatures of the surfaces become equilibrated. Hence, research would require an estimation of the period of time, which would guarantee equilibration with the environmental temperature. The application of both temperature and RH sensors within such a small space makes it possible to recognise the difference in equilibration and therefore see the RH change ahead of the temperature change. Lower sensor temperatures will result in the same proportion of water vapour in the air being interpreted as a higher RH, which was clearly shown by the heating trials (Figure 4). The faster the approach speed between the sensor and temperature source, the larger

RH output generated (artefact: Figures 6 and 8), probably as a result of the rather instantaneous presentation of skin surface water vapour and the more delayed transmission of skin associated temperature changes (Figure 7). The human body is a complicated thermoregulatory system that keeps core body temperature at approximately 37 °C through various mechanisms, including sweat evaporation [26]. In addition, choice of clothing material can be used to either enhance or hinder both heat and water vapour transfer out from the body, and thus by extension, will affect transfer to and from any surface the skin comes into contact with [27] and should be carefully considered for any experimental assessment of this interface, along with the surface RH characteristics [13,16]. This finding highlights the importance of not only building an integrated microenvironment (both temperature and humidity generator) simulation system but considering the impact of loading rate (e.g., sitting speed) when using either a human or a dummy to investigate RH changes at the body-seat interface [13].

In the heating trial, the RH reached equilibrium (Figure 4) after the water temperature (19.9 °C) in the tank had nearly returned to room temperature (22.5 °C ± 0.2 °C). This phenomenon supports the delay in the interpretation of RH by the system caused by the sensor failing to equilibrate quickly enough to changes in temperature. Hence, there will always be a prolonged period to achieve an accurate RH equilibrium (i.e., reading the accurate RH at the set temperature). The rate of change was found to be 0.9% RH/min for the whole testing process (from heating to cooling). These findings indicate that thermal conditions should be taken into account when analysing the RH variations [16] as the amount of water vapour present could also be expected to affect temperature transfer rate. This aspect has been verified with the calculations of HI which reflects the combined influence of temperature and RH (Figure 5). The rising and falling trend of calculated HI and measured temperature showed a similar pattern (difference in mean ± SD: 1.1 °C ± 0.5 °C). In addition, there was a strong negative correlation between temperature and RH. It is this association between temperature and RH that may underpin the "spike" phenomenon when measuring RH at the body-seat interface during the initial period of sitting (Figure 7). The DP tests showed that there were no obvious differences in the calculated values and previously published ones (technical bulletin Number 10, Lamtec Co., Bethel, PA, USA).

The general finding of a lack of difference between left mid-thigh and right mid-thigh is consistent with our previous reports showing the RH distribution to exhibit a symmetrical pattern when a healthy participant uses a standardised sitting posture (e.g. sitting upright without fidgeting or movement) [18]. As the thighs and coccyx are significantly different, it supports the necessity of utilising multiple sensors measurement points when investigating microenvironment characteristics over the whole contact surface. The slower rate of approach between the sensor and person sitting showed differences; however, the faster speed of approach failed to show a similar pattern (no significant difference among different measurement locations). This may be due to larger magnitude of air movement resulting in some water vapour being expelled from the small region where the sensors were placed during faster sitting process. This supports the need to pre-determine the rate of sitting and movement when deciding the protocol, as rapidity of change in position can differentially affect the size of any RH change associated with the movement.

4.3. Limitations

Though some apparently meaningful findings have been reported here, there were several limitations to the current study. Regarding the response time of sensors, we relied on the datasheet and previously published research results [18,19] when deciding which sampling frequency to choose. Though the typical response for the HTU21D is 5 s (Max = 10 s), it would be better to directly evaluate this performance in future work, to avoid the issues identified here related to the delay in equilibration of temperature. Then, the trials of sitting speed (either sitting down or sensor movement) were conducted consecutively (i.e., 10 repetitions at the same rate of movement), based on the consideration that the reliability and consistency between each sit down phase would more likely be higher if all the repeats of a single sitting speed were performed consecutively. However, selecting the speed randomly may have been a better solution to reduce additional issues (e.g., anxiety/boredom) associated with continuously repeating the activity at the same speed. Additionally, we assumed the movement to be linear, which may have added a slight variance based on choice of sitting movement method chosen by the subject for each test. A further limitation might have been the start/stop points of the stepping motor in the current study, which was determined by estimating the running time (assuming the distance and speed were constant values). However, it is difficult to ensure every movement was a uniform rectilinear motion due to the probably presence of mechanical and electronic errors. For more accurate measurements, a proximity sensor would have been a better solution to control the stop point of the stepping motor and allow more accurate determination of the end-point distances in this enclosed space. It might also be useful to repeat the study with more participants in order to verify the universality of the findings.

Notwithstanding the aforementioned weaknesses, the strengths of this study appear obvious:

1. Evaluate the performance of the temperature-humidity-integrated sensor and determine the potential for (and confounding factors underpinning) the artefact based changes in RH. Our results suggest that the HTU21D could be considered a more ideal choice for simultaneously measuring the microenvironment (both temperature and RH) changes at the body-seat interface.
2. Demonstrate that a rapid heating or cooling could have a strong impact on reported RH values owing to the environmental changes (such as thermal exchange) within a small area. It must be remembered that the body-seat interface RH has an association with the body temperature transmitted from user to sensors. As heat conduction through air is slow, the RH estimation at skin levels will be subject to artefact enhancement until the temperature of the sensor approximates that of the skin.
3. The initial sitting contact induced RH peak could be considered an artefact resulting from the increased moisture associated with the warmer body entering the small region over a colder sensor. This finding further highlights the importance of monitoring temperature changes while investigating the RH variations at the contact surface. This monitoring is not only important for the start of sitting, but also during prolonged periods of sitting as the person starts to fidget.

5. Conclusions

The findings support the hypothesis that the transient increase in RH at the onset of sitting is an artefact as a result of moisture from a warmer environment interacting with a colder sensor. This spike in RH occurred during the sensor movement trials (Section 2.5.1) and attained 80.2% $\pm$ 3.2% RH with a body-seat interface temperature of 29.4 °C $\pm$ 0.3 °C. Following the initial point of contact, the RH peak declined in magnitude as the temperature of the sensor increased up to the point of thermal equilibrium in that environment. The stable outcome of RH in the sensor movement trials was 65.5% $\pm$ 3.3% RH and temperature between skin surface and seat contact surface reached 32.4 °C $\pm$ 0.4 °C. Therefore, when evaluating microenvironment variations between the body and seat interface, it appears critically important for researchers to consider adapting methodological changes to limit the impact of movements when sitting and recording RH. Although RH changes are most noticeable in the initial contact period (approximately the first 40-seconds of data), it is likely that similar effects occur as the result of movements in prolonged sitting experiments. In addition, the correlation between RH and temperature is so strong that it is necessary to monitor temperature while investigating RH changes at the contact surface.

Author Contributions: P.W.M. and V.C. designed the experiments. Z.L. and J.L. performed the trials. M.L. analysed the data. All authors discussed the manuscript and engaged in the writing process.

Funding: This work was supported by the Natural Science Foundation of Heilongjiang Province (Grant No. F201421), the Harbin Scientific Innovation Project for Elite Young Researcher (Grant No. 2013RFQXJ093) and the Scientific Research and Talent Project of Education Department of Heilongjiang Province (Grant No.12541109, 12541140).

Conflicts of Interest: The authors declare no conflict of interest.

References

1. Coenen, P.; Gilson, N.; Healy, G.N.; Dunstan, D.W.; Straker, L.M. A qualitative review of existing national and international occupational safety and health policies relating to occupational sedentary behavior. *Appl. Ergon.* **2017**, *60*, 320–333. [CrossRef]
2. Colley, R.C.; Garriguet, D.; Janssen, I.; Craig, C.L.; Clarke, J.; Tremblay, M.S. Physical activity of Canadian adults: Accelerometer results from the 2007 to 2009 Canadian Health Measures Survey. *Health Rep.* **2011**, *22*, 7–14.
3. Healy, G.N.; Matthews, C.E.; Dunstan, D.W.; Winkler, E.H.; Owen, N. Sedentary time and cardio-metabolic biomarkers in US adults: NHANES 2003–2006. *Eur. Heart J.* **2011**, *32*, 590–597. [CrossRef] [PubMed]
4. Matthews, C.E.; Chen, K.Y.; Freedson, P.S.; Buchowski, M.S.; Beech, B.M. Amount of time spent in sedentary behaviors in the United States, 2003–2004. *Am. J. Epidemiol.* **2008**, *167*, 875–881. [CrossRef] [PubMed]
5. Hamilton, M.T.; Hamilton, D.G.; Zderic, T.W. Role of low energy expenditure and sitting in obesity, metabolic syndrome, type 2 diabetes, and cardiovascular disease. *Diabetes* **2007**, *56*, 2655–2667. [CrossRef] [PubMed]
6. Chai, C.Y.; Sadou, O.; Worsley, P.R.; Bader, D.L. Pressure signatures can influence tissue response for individuals supported on an alternating pressure mattress. *J. Tissue Viability* **2017**, *26*, 180. [CrossRef] [PubMed]
7. Redelings, M.D.; Lee, N.E.; Sorvillo, F. Pressure ulcers: More lethal than we thought? *Adv. Skin Wound Care* **2005**, *18*, 367–372. [CrossRef]
8. Marchione, F.G.; Araújo, L.Q.; Araújo, L.V. Approaches that use software to support the prevention of pressure ulcer: A systematic review. *Int. J. Med. Inform.* **2015**, *84*, 725–736. [CrossRef]
9. Gefen, A. How do microclimate factors affect the risk for superficial pressure ulcers: A mathematical modeling study. *J. Tissue Viability* **2011**, *20*, 81–88. [CrossRef]
10. Irzmanska, E.; Lipp, B.; Kujawa, J.; Irzmanski, R. Textiles preventing skin damage. *Fibres Text. East. Eur.* **2010**, *18*, 84–90.
11. Irzmanska, E.; Charlusz, M.; Kujawa, J.; Kowalski, J.; Pawlicki, L.; Irzmanski, R. Using impedance plethysmography to evaluate antidecubital underlay systems for chronically immobilized patients. *Adv. Clin. Exp. Med.* **2010**, *19*, 637–651.
12. Magalhaes, M.G.; Gragnani, A.; Veiga, D.F.; Blanes, L.; Galhardo, V.C.; Kallas, H.; Juliano, Y.; Ferreira, L.M. Risk Factors for pressure ulcers in hospitalized elderly without significant cognitive impairment. *Wounds-Compend. Clin. Res. Pract.* **2007**, *19*, 20.
13. Freeto, T.; Cypress, A.; Amalraj, S.; Yusufishaq, M.S.; Bogie, K.M. Development of a sitting microenvironment simulator for wheelchair cushion assessment. *J. Tissue Viability* **2016**, *25*, 175–179. [CrossRef]
14. Zitek, P.; Vyhlidal, T.; Simeunovic, G.; Novakova, L.; Cizek, J. Novel personalized and humidified air supply for airliner passengers. *Build. Environ.* **2010**, *45*, 2345–2353. [CrossRef]
15. Conceicao, E.E.; Lucio, M.R.; Rosa, S.P.; Custodio, A.V.; Andrade, R.L.; Meira, M.A. Evaluation of comfort level in desks equipped with two personalized ventilation systems in slightly warm environments. *Build. Environ.* **2010**, *45*, 601–609. [CrossRef]
16. Ferguson, M.; Hirose, H.; Nicholson, G.; Call, E. Thermodynamic rigid cushion loading indenter: A buttock-shaped temperature and humidity measurement system for cushioning surfaces under anatomical compression conditions. *J. Rehabil. Res. Dev.* **2009**, *46*, 945–956. [CrossRef]
17. Diesing, P.; Hochmann, D.; Boenick, U.; Kraft, M. A novel method for patient-oriented assignment of wheelchair cushions based on standardized laboratory testing procedures. *Biomed. Eng.* **2005**, *50*, 188–194. [CrossRef]
18. McCarthy, P.W.; Liu, Z.F.; Heusch, A.I.; Cascioli, V. Assessment of humidity and temperature sensors and their application to seating. *J. Med. Eng. Technol.* **2009**, *33*, 449–453. [CrossRef] [PubMed]
19. Liu, Z.F.; Cheng, H.F.; Luo, Z.M.; Cascioli, V.; Heusch, A.I.; Nair, N.R.; McCarthy, P.W. Performance assessment of a humidity measurement system and its use to evaluate moisture characteristics of wheelchair cushions at the user–seat interface. *Sensors* **2017**, *17*, 775. [CrossRef]

Sensors **2019**, *19*, 1471

20. Liu, Z.F.; Chang, L.; Luo, Z.M.; Cascioli, V.; Heusch, A.I.; McCarthy, P.W. Design and development of a thermal imaging system based on a temperature sensor array for temperature measurements of enclosed surfaces and its use at the body-seat interface. *Measurement* **2017**, *104*, 123–131. [CrossRef]

21. Zemp, R.; Taylor, W.R.; Lorenzetti, S. Seat pan and backrest pressure distribution while sitting in office chairs. *Appl. Ergon.* **2016**, *53*, 1–9. [CrossRef]

22. Meilinger, T.; Garsoffky, B.; Schwan, S. A catch-up illusion arising from a distance-dependent perception bias in judging relative movement. *Sci. Rep.* **2017**, *7*, 17037. [CrossRef] [PubMed]

23. Morrison, S.; Sosnoff, J.J.; Heffernan, K.S.; Jae, S.Y.; Bo, F. Aging, hypertension and physiological tremor: The contribution of the cardioballistic impulse to tremorgenesis in older adults. *J. Neurol. Sci.* **2013**, *326*, 68–74. [CrossRef] [PubMed]

24. Anderson, G.B.; Bell, M.L.; Peng, R.D. Methods to Calculate the Heat Index as an Exposure Metric in Environmental Health Research. *Environ. Health Perspect.* **2013**, *121*, 1111–1119. [CrossRef] [PubMed]

25. Alduchov, O.A.; Eskridge, R.E. Improved Magnus' form approximation of saturation vapor pressure. *J. Appl. Meteorol.* **1996**, *35*, 601–609. [CrossRef]

26. Stancic, M.; Kasikovic, N.; Grujic, D.; Novakovic, D.; Milosevic, R.; Ruzicic, B.; Gersak, J. Mathematical models for water vapour resistance prediction of printed garments. *Color Technol.* **2018**, *134*, 82–88. [CrossRef]

27. Li, F.Z.; Li, Y. Effect of clothing material on thermal responses of the human body. *Model. Simul. Mater. Sci. Eng.* **2005**, *13*, 809–827. [CrossRef]

sensors

MDPI

Article

A Laboratory Study on Non-Invasive Soil Water Content Estimation Using Capacitive Based Sensors

Amir Orangi, Guillermo A. Narsilio and Dongryeol Ryu

Department of Infrastructure Engineering, School of Engineering, The University of Melbourne, Parkville, VIC 3010, Australia; amir.orangi@unimelb.edu.au (A.O.); dryu@unimelb.edu.au (D.R.)
* Correspondence: narsilio@unimelb.edu.au; Tel.: +61-3-8344-4659

Received: 1 October 2018; Accepted: 19 November 2018; Published: 5 February 2019

Abstract: Soil water content is an important parameter in many engineering, agricultural and environmental applications. In practice, there exists a need to measure this parameter rather frequently in both time and space. However, common measurement techniques are typically invasive, time-consuming and labour-intensive, or rely on potentially risky (although highly regulated) nuclear-based methods, making frequent measurements of soil water content impractical. Here we investigate in the laboratory the effectiveness of four new low-cost non-invasive sensors to estimate the soil water content of a range of soil types. While the results of each of the four sensors are promising, one of the sensors, herein called the "AOGAN" sensor, exhibits superior performance, as it was designed based on combining the best geometrical and electronic features of the other three sensors. The performance of the sensors is, however, influenced by the quality of the sensor-soil coupling and the soil surface roughness. Accuracy was found to be within 5% of volumetric water content, considered sufficient to enable higher spatiotemporal resolution contrast for mapping of soil water content.

Keywords: agriculture; capacitive sensors; dielectric constant; remote sensing; surface soil water content

---

1. Introduction

Soil water content is a parameter with implications in an array of engineering, hydrology, climate science, water resource management, remote sensing and agricultural applications [1–6]. The challenge of increasing water use in agriculture, which is known to be the largest consumer of water resources (e.g., see [7]), can be alleviated by better-informed irrigation decisions and smart farming systems that are based on accurate measurements of soil water content [8–10]. In addition, accurate and rapid measurements of soil water content can enhance site assessments in a broad range of civil engineering applications such as road construction, since the soil moisture is an important parameter to derive the strength and the integrity of the infrastructure [11]. Furthermore, in bushfire management, the fuel availability estimates used for issuing warnings are partly based on the soil moisture deficit [12].

Surface soil moisture comprises only 0.05% of the Earth's total fresh water. Although this value is small, the amount of soil moisture is imperative in agriculture for crop development, irrigation management, crop type selection and plant stress [13–16]. Additionally, spatiotemporal variations of near surface soil water content are of paramount significance in a number of applications due to its very large inhomogeneity [17–19]. As such, the near surface soil water content is an integral hydrological and meteorological parameter for ground truthing remotely sensed data, mapping variable sources of streamflow, developing large scale surface water and energy balance models and improving the land component of climate models, including global circulation models [17,20–26]. In precision agriculture, it is the soil water content in the root zone, not the near surface water content, that determines the amount of water available to a plant. However, the water available to the plant often can be inferred from near surface soil water content information (e.g., see [19,27–30]). Nonetheless, with the current soil water content measurement techniques measuring the surface soil water content become difficult to assess its spatiotemporal variability [31,32].

Soil water content can be directly measured using the oven drying method which is accurate and inexpensive; however, it is time-consuming and labour-intensive. In addition, there are indirect techniques which utilise other soil parameters as a proxy to estimate soil water content. Neutron probes are commonly used for these indirect techniques; however, there are limitations associated with their use. These limitations are primarily due to the probes containing radioactive materials and include the high cost of equipment, the requirement of a certificate to operate, the inability to use as a continuous monitoring tool and unreliability to estimate near surface soil water content [2,20,33]. Furthermore, the common methods of measuring soil water content often cannot provide immediate feedback [34]. The disadvantages of traditional soil water measurement methods associated with time and cost are exacerbated by the large spatial extent of measurements required for irrigation management in agriculture and motivate the development of cost-effective and non-invasive alternatives [35,36].

Alternative techniques which address some of the limitations of traditional methods include dielectric methods. In these, the electrical properties of soil are utilised as a proxy to estimate soil water content. For example, Time Domain Reflectometry (TDR) probes, Frequency Domain Reflectometry (FDR) probes, capacitive probes, impedance probes, Ground Penetration Radar (GPR) and Electromagnetic conductivity (EM) antennas have been used to estimate water content in various applications (e.g., see [10,37–41]). However, the accessibility of these methods is limited by the high cost of equipment and difficult result interpretation. Furthermore, despite the non-invasive nature of GPR and EM antennas, the requirement of probe insertion into the soil makes the TDR, FDR, capacitive and impedance probes labour intensive, particularly for hard or dense soils. Additionally, by being invasive, repeated measurements at the same location can make the measurements unreliable [42]. To compensate for the high cost involved in some of the aforementioned methods, there exist other common low-cost sensors which, although invasive, demonstrate good performance. The heat pulse soil moisture sensors using single or dual probe designs have been introduced in this context [43–50]. Furthermore, recent developments in capacitive soil moisture sensors have enabled low cost means in soil water content measurement [49,50]. Nonetheless, they have not eliminated the need for probe insertion. For this purpose, a needle-free heat pulse sensor system has recently been developed [51]. Although this sensor has no needle and is inexpensive, burying it in the soil causes soil disturbance [51]. None of these developments, however, have focused on non-invasive measurements of near surface soil water content. Instead, remote sensing applications have been widely used as the primary source of information of surface soil water content. However, they often lack the required resolution for certain applications [31,52]. It has been suggested that understanding the sub-footprint scale of the variability of remotely sensed soil water content is an important factor to fully utilise these data [31].

Motivated by the importance of soil water content and the drawbacks of its current measurement techniques, particularly for near surface soil water content, this research aims to develop a new non-invasive, low cost and capacitive-based technique for estimating near surface soil water content. We hypothesise that given the accuracy of the relationship between soil water content and its dielectric properties, which is widely known, the surface soil water content can be estimated from the surface using a non-invasive capacitive sensor (since capacitance is linked to the dielectric constant and the sensor geometry). To evaluate this hypothesis, we first compared three new non-invasive capacitive sensors developed to estimate soil volumetric water content and we examined their performance for four different soil types and for a range of water contents. Subsequently, based on the comparative performance of these three sensors, a fourth sensor was designed and manufactured to substantially reduce the limitations of the previous versions. The sensors, particularly the fourth one, demonstrated great potential in detecting variation in soil water content from the ground surface. Moreover, it is concluded that the soil-sensor coupling and roughness of the soil sample surface play an important influencing role in the performance of the sensors.

To address the objective of this research, this manuscript is organised as follows: In Section 2, an overview of dielectric permittivity and soil moisture is introduced. Sections 3 and 4 comprise descriptions of the materials and methodologies used. The results, analyses and discussion are presented in Section 5. In Section 6, we discuss potential practical applications of the sensors as well as the limitations of the research. Finally, summaries of the findings and recommendations for future work are presented in Section 7.

2. Theory and Background: Dielectric Permittivity and Soil Moisture

Current non-destructive soil water content estimation techniques such as GPR and TDR are based on measuring the dielectric permittivity of soil. The dielectric permittivity, ε (F/m), is a complex number which measures the degree to which a material is polarised when it is subjected to an electrical field and it can be represented as shown by [41]:

$$\varepsilon = \varepsilon' - j\varepsilon'' \tag{1}$$

where ε' is the real component of the dielectric permittivity (F/m), $j = \sqrt{-1}$ is the imaginary number and ε'' is the imaginary component of the dielectric permittivity (F/m) known as the dielectric loss. The ratio between a material's dielectric permittivity and that of air ($\varepsilon_o \approx 8.85 \times 10^{-12}$ F/m) is known as the relative dielectric permittivity, κ and can be written as:

$$\kappa = \frac{\varepsilon}{\varepsilon_o} = \kappa' - j\kappa'' \tag{2}$$

where κ' is known as the dielectric constant and κ'' is known as the loss factor. In an unsaturated soil, water has the highest dielectric constant ($\kappa' \cong 80$), which is noticeably larger than the dielectric constant of minerals ($2 < \kappa' < 7$) and of air ($\kappa' = 1$). The bulk dielectric permittivity of soil, a mixture of these three elements is, therefore, influenced mostly by the water content. Indeed, a strong correlation between the soil volumetric water content and the (real) dielectric constant is reported in the literature. There are several studies which investigated the correlation between soil volumetric water content and its real dielectric constant, considering parameters such as soil type, salinity, density, temperature and frequency of measurements (e.g., [53–57]). Within this context, Topp, Davis and Annan [53] performed TDR measurements on four types of soil to propose what it is today the most commonly used empirical model: for a low-loss homogenous material (i.e., low or negligible κ''), the correlation between the apparent dielectric constant, κ' and its volumetric water content, θ, is:

$$\theta = 4.3 \times 10^{-6} \kappa'^3 - 0.00055 \kappa'^2 + 0.0292\kappa' - 0.053 \tag{3}$$

Further, capacitance, C, is the ability of a material to store an electrical charge. The capacitance of a capacitor is related to the dielectric constant of the dielectric material used as the insulator [58] such that:

$$C = \kappa g \varepsilon_0 \tag{4}$$

where κ is the relative dielectric permittivity, g is a geometric constant and ε_0 is the permittivity of a vacuum (F/m). Measurement of soil permittivity through capacitance methods was first introduced by Dean, et al. [59], who developed a capacitance sensor operating at a frequency of 150 MHz for the purpose of creating a cost-effective and safe in situ method [58]. Furthermore, due to their relatively low cost and ease of operation, capacitive sensors are becoming increasingly popular among researchers and practitioners [60]. Most importantly, the relationship between water content and the dielectric constant of a soil is widely accepted to be accurate [42,58]. We compare four new capacitive sensors developed to non-invasively estimate the soil volumetric water content by utilising the relationship between the volumetric soil water content and the dielectric constant (Equation (3)) and the relationship between the capacitance and the dielectric constant (Equation (4)).

3. Materials and Methods

This section describes the experimental framework including the material and methods used to address the objectives of this research.

3.1. Tested Soils

Four different soil types collected from Victoria, Australia, whose characteristics are summarised in Table 1, are used for testing. Soils are selected to cover a range of grain sizes and textures based on grain size distribution analysis according to Australian Standards [61,62]. In addition, the plasticity index (Plastic Limit, PL and Liquid Limit, LL) were determined following Australian Standards [63,64]. The Organic Matter (OM) was measured using the Loss on Ignition (LOI) method described in the American Society for Testing and Materials (ASTM) standards [65]. The salinity of the samples was estimated using the conductivity, σ, of the samples at saturation point ($\sigma = \kappa'' \cdot \omega \cdot \varepsilon_0$), where κ'' is the loss factor (see Equation (2)) and ω is the angular frequency, as proposed by Santamarina and Fam as well as Narsilio, et al. [66,67].

Table 1. Characterisation of soil samples.

Soil Sample	Location	Clay (%)	Silt (%)	Sand (%)	PL (%)	LL (%)	OM (%)	Salinity (dS/m)	USCS Symbol
Brighton Group Sand	Brighton, VIC	<1	2	97	NA	NA	0.08	1.5	SP
Silty Sand	Melbourne, VIC	<1	14	85	NA	NA	0.37	2.3	SM
Silty Clay (Camb)	Camberwell, VIC	48	42	10	19	28	0.18	3.0	CL
Clayey Silt (Bun)	Buninyong, VIC	13	70	17	30	39	0.33	1.6	ML

PL—plastic limit; LL—liquid limit; OM—organic matter; SP—poorly graded sand; SM—fine-grained silty sand; CL—low to medium plasticity silty clay; ML—low to medium plasticity clayey silt.

Based on the unified soil classification system (USCS) described in the Australian Geotechnical Site Investigations standard [68], the first soil is classified as a poorly graded sand (SP) denoted in this manuscript (and locally known) as Brighton Group Sand; the second soil as a fine-grained silty sand (SM) referred to as Silty Sand in this manuscript; the third soil as a low to medium plasticity silty clay sample (CL) denoted as Camb Clay in this manuscript and the fourth sample is classified as a low to medium plasticity clayey silt (ML) referred to as Bun Silt in this manuscript. These clay samples belong to the Silurian Melbourne geological formation, which contains primarily illite and kaolinite minerals. Based on the estimated values of salinity, the Brighton Group Sand and the Bun Silt are considered as non-saline soils whereas the Silty Sand and Camb Clay samples are considered as moderately saline according to Agriculture Victoria [69].

3.2. Dielectric Probe: Benchmark Dielectric Measurements

The complex dielectric properties of soil samples can be measured by means of an open-ended coaxial line technique [70]. In this work, a 2.2 mm diameter coaxial slim form Agilent dielectric probe (Keysight Technologies, Santa Rosa, CA, USA) (with a 0.51 mm diameter centre conductor and a 1.68 mm diameter insulator) connected to an N9923A FieldFox Vector Network Analyser (VNA) (Keysight Technologies), was utilised to measure the complex dielectric properties of the soils at different frequencies. These measurements help in evaluating the performance of the capacitive sensors. The slim form probe is reported in the literature to have been used to investigate the relationships between soil dielectric properties and other parameters such as water content, thermal conductivity, temperature, frequency and pH [71–73]. The complex scattering parameter of the material under test is measured and subsequently converted to the complex dielectric permittivity by means of proprietary software [74]. All the experiments were conducted in a controlled laboratory environment with a constant temperature range of 19 to 21 degrees Celsius.

3.3. Capacitive Sensors

The four capacitive sensors used in this study are depicted in Figure 1 alongside an Arduino-based board use as a controller board. The sensors are (1) an AD7746 sensor, denoted as "Circular," (2) an MPR121 sensor, denoted as "Rectangular," (3) a PCB Gadget sensor, denoted as "PCB" and (4) a newly designed and built sensor denoted as "AOGAN." The sensors are Capacitance-to-Digital Converters (CDC). The Circular and Rectangular sensors are typically used as keypads; however, in this work, they were programmed to measure the capacitance of the soil samples. The Circular, Rectangular and AOGAN sensors were connected to an Arduino-based board (Freetronics, Croydon South, VIC, Australia) and utilised a C++ platform to communicate and transmit the measured capacitance values. Similarly, the PCB sensor transmitted the capacitance reading through a USB cable to a CoolTerm computer program (provided by PCB Gadget) without the need of an Arduino-based controller board.

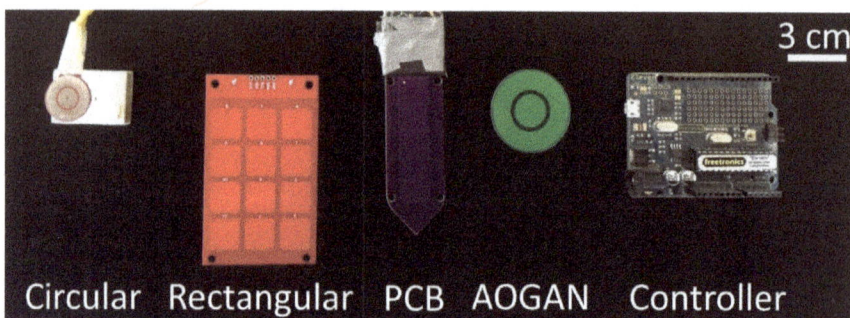

Figure 1. Capacitive sensors: Circular, Rectangular, PCB Gadget (PCB) and AOGAN sensors (left to right). The Arduino-based board controller is also shown to the right.

Regarding sensor specifications, the Circular sensor is composed of two concentric plates comprising the electrodes. This sensor can measure up to 24 pF capacitance, with a linearity of $\pm 0.01\%$ and accuracy of ± 4 fF factory calibrated. The positive supply voltage can vary between + 0.3 and + 6.5 V and it has an operational frequency of approximately 32 kHz [75]. The Rectangular sensor has 12 capacitance sensing inputs and was programmed to measure a capacitance range from 0.45 pF to over 340 pF (depending on the programming code), has a positive supply voltage of 1.71 to 3.6 V operated at 400 kHz [76] and the sensing electrodes are covered with an insulating layer. The PCB sensor is a capacitive sensor comprising a single electrode to measure changes in the capacitance of a material operating at a frequency of 500 kHz [77]. The fourth sensor, the AOGAN sensor, was manufactured by adopting a similar shape to the Circular sensor, with an insulating layer, with a similar controller board as the Rectangular sensor and with an operating frequency of 400 kHz. The sensor was also operated by an Arduino-based board.

4. Experimental Procedure

A description of the testing and development of the sensors for the non-invasive soil water content estimation is presented in the following sections.

4.1. Sample Preparation and Dielectric Measurements

Soil samples were crushed and subsequently prepared from the air-dry condition to saturation, by incrementally adding deionised water. This incremental addition of water was achieved by thoroughly mixing the soil and allowing adequate curing time for the samples to attain a homogeneous state. Deionised water was used to minimise the introduction of any foreign ions to the soil samples, which may have potentially influenced the dielectric properties. The soil was then transferred to a non-dielectric (PVC) container with a known volume and a size adequate to accommodate the Agilent dielectric probe and capacitive sensor. We initially used a PVC container with similar dimensions to a standard compaction mould (approximate volume of 1000 cm^3, see [78]) and later changed to a smaller PVC mould with the same diameter (10 cm) and a volume of approximately 160 cm^2. We initially tested the bulk density effects, by preparing samples at different dry densities using the larger mould. However, since the sensitivity of the sensor to density variation was deemed to be insignificant, we opted to use a smaller mould to expedite the experimental program. It is important to note that the size of the smaller PVC mould was selected such a way that the thickness of the soil was larger than the sensing sphere and the sensing geometrical element of the sensor(s) and the open-ended coaxial probe. For each dielectric and capacitance measurement, the container volume and the mass of the soil were recorded, to be used in the computation of the sample's volumetric water content. The PVC containers were chosen to minimise electromagnetic interference. Prior to the dielectric measurements, the probe was calibrated against air, a shorting block and deionised water. Thereafter, at least three measurements were taken for a given sample, on different parts of the sample's surface area to ensure that the dielectric constant measurement was an accurate representation of the entire sample. Contact between the probe and the soil was carefully maintained to ensure that there was no air trapped between them. Figure 2 depicts a typical instrument and sample setup used in the experiment.

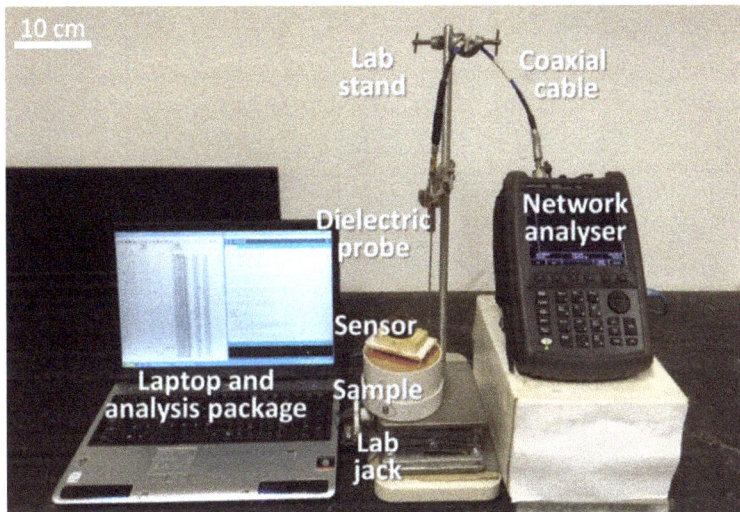

Figure 2. Typical experimental setup.

The dielectric probe measurements were followed by measurements using the four capacitive sensors (explained in detail in the next section). Lastly, a sub-sample was retrieved for subsequent gravimetric and volumetric water content, θ, calculations, using the soil sample dry density derived from the known volume of the container and the measured soil mass [79]. It is worth mentioning that the sub-samples for oven drying were retrieved from the uppermost layer of the sample (around 10 mm) which was estimated to be within the sensing volume of the various probes used. This was to ensure that the sensor outputs were calibrated against a representative volume. The approach undertaken to conduct the capacitive measurements is described in the next section.

4.2. Capacitive Sensor Measurements

Once the sample was prepared, the following measurement protocol was followed for each of the sensors. Firstly, a measurement was conducted whilst the sensor was free in the air and recorded as an air measurement. Subsequent measurements were undertaken by placing the sensor against the surface of the sample and by applying a weight (minimum 200 g) on top of the sensors to ensure a good soil-sensor contact was maintained, without any noticeable air gap between the soil and the sensor. Air gaps could potentially lead to errors in the capacitive reading and thus in the soil water content estimation. To ensure the soil-sensor contact was maintained, a slightly heavier load was used for sensors with a larger footprint. Once the full contact was maintained and consistent readings were obtained from the sensor, the data was recorded on a computer. This procedure was repeated typically three times for each sample to obtain readings that were accurate representations of the entire sample. Once the air and sample readings were recorded, the air reading was subtracted from the reading taken from the sample. This was done to minimise the effect of the environment (such as humidity and temperature) on the measurements. Moreover, this can be considered as a basic and simple calibration to normalise the measurements with respect to the air reading.

It is worthwhile to note that the effect of pressure on the output of the sensor was tested through a separate set of experiments. The sensor readings were monitored while various pressures were applied to the sensor as it was sitting on a flat surface. Once full contact was maintained between the sensor and the material under test, changing the amount of pressure was proven to have no impact on the readings (results are not shown).

The approach for the Rectangular sensor was slightly different due to its multiple-electrode design. As this sensor is comprised of twelve sensing electrodes, twelve readings were obtained from a 'single' measurement. In addition, due to the use of 12 electrodes and the relatively larger size of the sensor unit compared to the Circular sensor, some variations were observed in the readings (results are shown later in the paper in Sections 5.2 and 5.7). This is likely due to the fact that despite the sample surface being levelled, there still existed some relative surface roughness which may have created an uneven contact between the soil and some of the electrodes, causing noticeable dissimilarities between different electrode readings. To overcome this issue, instead of averaging the twelve readings, which was the approach adopted in the work previously reported by Orangi, et al. [80], the maximum reading among the twelve readings for a given measurement was used for the analysis. Based on the observations throughout this work and the findings highlighted by Orangi, Withers, Langley and Narsilio [80], it was assumed that for a given sample, the larger reading values were derived from the better soil-sensor contact conditions and thus more representative of the true soil water condition. Further details are included in the following section.

With regard to the PCB sensor, the measurements were conducted without using a weight; the sensor was simply held by hand against the sample surface whilst ensuring full contact was maintained. The measurements with the Circular and AOGAN sensors were conducted as described in the general procedure.

5. Results, Analyses and Discussion

Firstly, the results are summarised for each sensor and soil sample and individual calibrations are derived. Next, an evaluation of the applicability of a single calibration (all soil types) for each sensor is presented. This is followed by an evaluation of the efficacy of using a separate calibration for sandy soils (i.e., combining the Brighton Group Sand and Silty Sand data—referred to as sand group in this manuscript) and for cohesive soils (i.e., combining Camb Clay and Bun Silt data—referred to as clay groups in the manuscript) for each sensor. The dielectric properties of the samples are then estimated based on capacitive measurements and lastly, the effect of soil sensor coupling and surface roughness on sensor performance is evaluated by using the results of the present study and of Orangi, Withers, Langley and Narsilio [80].

5.1. Circular Sensor

Results of the Circular sensor performance against volumetric water content are shown in Figure 3. Each plot includes capacitance readings measured by the sensor and the dielectric constants from the dielectric probe versus the volumetric water content for each soil sample. The capacitance and dielectric measurement data are plotted with blue squares and black triangular markers, respectively. Moreover, the "expected" volumetric water content based on the Topp calibration and the measured dielectric constant (Equation (4)) is superimposed on the plot (dashed grey trend line).

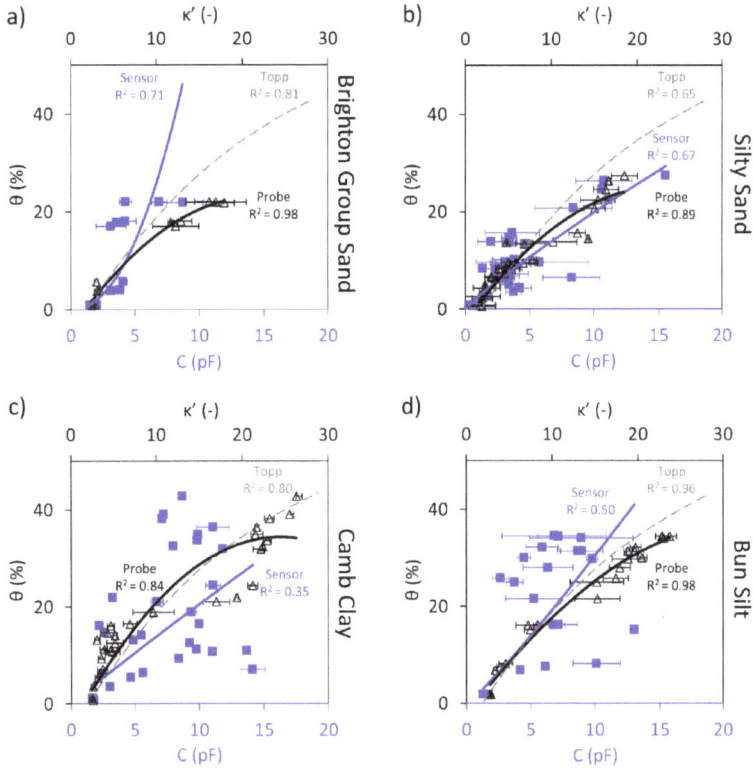

Figure 3. Circular sensor capacitance, C, readings (blue square markers and trendline) and dielectric constant, κ', measurements (black triangular markers and trendline) shown against the volumetric water content, θ; the Topp equation is also shown (dashed grey trendline). Shown for: (**a**) Brighton Group Sand (**b**) Silty Sand (**c**) Camb Clay and (**d**) Bun Silt.

Figure 3 illustrates an increasing trend captured by the Circular sensor for the four samples; however, with different levels of accuracy. As was explained in the theory and background section, an increase in the soil volumetric water content causes an increase in the dielectric constant of the soil, due to the larger number of water dipoles. An increase in the capacitance is also expected since Equation (4) shows that the relationship between capacitance and the dielectric constant is directly proportional.

For the Brighton Group Sand and the Silty Sand samples (i.e., the coarse-grained soil samples), the sensor was able to capture the variation of water content with capacitance; however, the correlations show errors in the order of 10%.

For the Camb Clay and Bun Silt (i.e., the fine-grained soil samples), it is observed that the Circular sensor is able to capture the increasing trend; however, the correlations compared to the sand samples are significantly less obvious, as shown by the significantly reduced coefficient of correlation, R^2.

The Topp calibration predictions shown by the dashed grey trendlines show a good agreement up to approximately 5% and 12% for the Brighton Group Sand and Silty Sand, respectively. However, as the water content increases, the Topp calibration overestimates the data. Surprisingly, the Topp calibration seems to fit the measured data for the cohesive soils tested here better than for the sandy soils, which is contrary to the assumptions made in deriving the calibration. It is worth mentioning that the R^2 for the Topp calibration describes how well this calibration captures the data.

Table 2 quantitatively summarises the best fit models and the measure of errors obtained for the Circular Sensor and the dielectric probe for the four soil samples. Additionally, the universal Topp calibration performance for each soil type is assessed. The results in the table confirm that the Circular Sensor can capture the increasing trend for the sand samples. Furthermore, for the cohesive samples, a very weak increasing trend could be identified from the data for both Camb and Bun samples, however, with an unacceptable level of accuracy and large errors. Despite the relatively low R^2 values for capacitive readings, it is worth mentioning that the dielectric probe measurements have shown some similar variations compared to the corresponding capacitive measurements for the first two soil samples.

Table 2. Summary of the Circular Sensor and dielectric probe performance against each soil sample.

Soil Sample	Circular Sensor			Dielectric Probe		
	Equation	R^2	RMSE (%)	Equation	R^2	RMSE (%)
Brighton Group Sand	$\theta = 0.43C^{2.16}$	0.71	9.71	$\theta = 0.057C^2 + 2.48C - 3.92$	0.98	1.08
Silty Sand	$\theta = 2.57C^{0.88}$	0.67	4.23	$\theta = -0.052C^2 + 2.4C - 2.5$	0.89	2.46
Silty Clay (Camb)	$\theta = 2.26C^{0.96}$	0.35	12.38	$\theta = 0.065C^2 + 3.19C - 4.5$	0.84	4.92
Clayey Silt (Bun)	$\theta = 2.31C^{1.12}$	0.50	12.15	$\theta = 0.034C^2 + 2.34C - 2.34$	0.98	1.80

θ = Volumetric Water Content, C = Capacitance (pF), R^2 = Coefficient of Correlation, RMSE = Root Mean Square Error.

A two-fold cross-validation analysis for the Circular Sensor was conducted (similar to other sensors in the next sections). In this analysis, half of the data is used for conducting a calibration and subsequently, the remainder of the data is used for validation. The results are summarised in Table 3 which further proves the weak performance of the Circular Sensor based on the low R^2 values and large errors.

Table 3. Cross validation analysis for the Circular sensor for each soil.

Soil Sample	Calibration		Validation	
	R^2	RMSE (%)	R^2	RMSE (%)
Brighton Group Sand	0.69	7.63	0.30	0.68
Silty Sand	0.63	4.25	0.62	4.82
Silty Clay (Camb)	0.31	11.82	0.27	13.69
Clayey Silt (Bun)	0.44	11.27	0.31	14.48

It is important to note that during measurement with this sensor, an instability issue of the reading occurred and the sensor could not retrieve any readings for some samples. This instability issue is attributed to the controller board of the Circular Sensor, as well as the lack of permanent insulating coating on the electrodes which may have created short circuits during measurements and the difficulties in achieving full soil-sensor coupling.

These reasons collectively have resulted in the relatively low performance of the Circular Sensor, particularly for fine-grained soils. Despite its weak performance, Circular Sensor demonstrates promising potential in detecting the variation in soil water content using non-invasive capacitive sensors. In view of the observed limitations, we have developed the Rectangular Sensor, whose results are described next.

5.2. Rectangular Sensor

The Rectangular sensor comprised 12 electrodes and each electrode returned a reading upon being in contact with the soil sample. Essentially, since the water content of the sample is envisaged to be homogeneous due to the sample preparation method, it is expected that the output from the 12 electrodes are similar or only with marginal variations due to electrodes layout. Therefore, it would be reasonable to report the mean of the 12 readings as the capacitive value for a given sample. This approach was adopted in a previous study by Orangi, Withers, Langley and Narsilio [80]; however, the Rectangular sensor performance was deemed unsatisfactory in that preceding study. In this work, the maximum reading (instead of the average) was adopted. The rationale for this choice was explained previously in Section 4.2 and the results are given in Section 5.7.

Figure 4 summaries the results of the Rectangular sensor against the volumetric water content. With the Brighton Group Sand and the Silty Sand samples, the (directly proportional) trend between the sensor readings and the volumetric water content can be clearly seen. In Figure 4a,b (coarse-grained soils), the sensor readings (blue square markers) show a strong correlation with increasing volumetric water content, which resembles the variations observed in the measured real dielectric constants (black triangular markers).

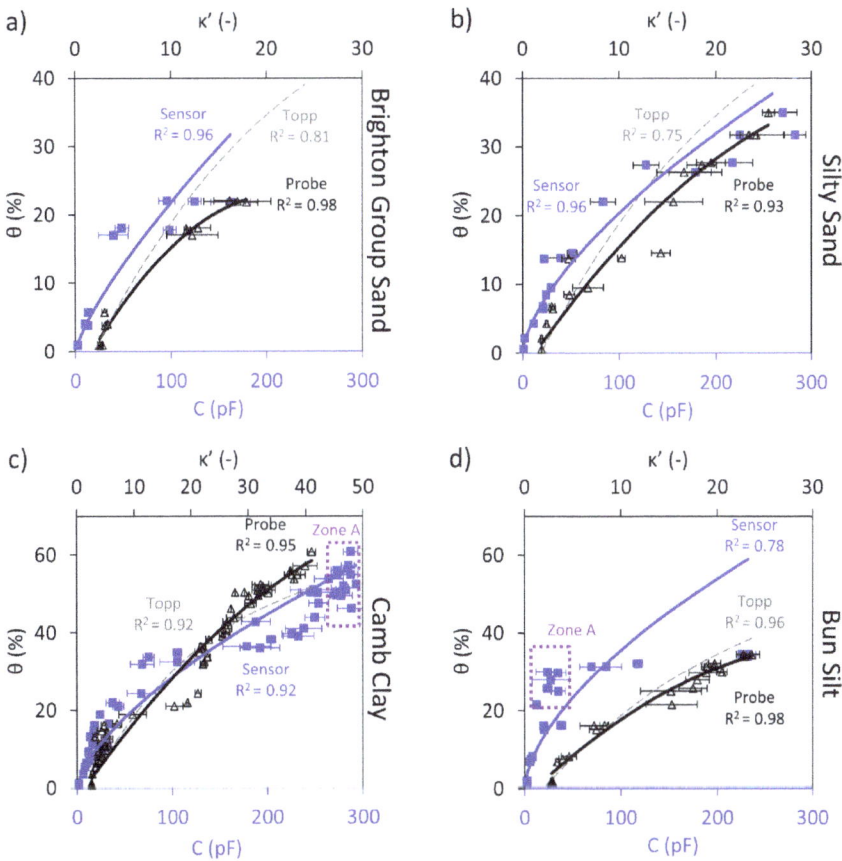

Figure 4. Rectangular sensor capacitance, C, readings (blue square markers and trendline) and dielectric constant, κ', measurements (black triangular markers and trendline) shown against the volumetric water content, θ; the Topp equation is also shown (dashed grey trendline). Shown for: (**a**) Brighton Group Sand (**b**) Silty Sand (**c**) Camb Clay and (**d**) Bun Silt.

For the fine-grained soils tested, the proportional trend between the capacitive reading and the volumetric water content can be clearly seen for both samples, as opposed to for the Circular sensor, where this trend had a weak resemblance due to the discussed limitations. Moreover, the instability of readings and the limited measurement range issues encountered by the Circular sensor are resolved here. The improvements are considered to be the result of the new controller as well as the larger geometry of the sensor, which facilitated maintaining full contact between the soil and sensor. However, it can be seen that electrodes returned almost constant readings despite the increase in sample water content (Zone A in Figure 4d), presumably due to the partial contact of electrodes for the Bun Silt samples with rougher soil surfaces. Indeed, soil trimming and surface smoothing were more difficult to achieve for the Bun Silt samples with water content between 20% and 30% (close to optimal water content). Otherwise, the trend is well captured by the Rectangular sensor for each of the sandy and cohesive samples.

The issue with the limited measurement range appeared for this sensor as well; however, at a much higher water content than when using the Circular sensor. As a result, the measured capacitive data beyond approximately 45% volumetric water content form a cluster of data points, as illustrated in Zone A of Figure 4c. That is, beyond approximately 40% water content, the sensor reached its upper limit and could no longer capture variation in the water content. Although the sensor was unable to differentiate the water content beyond this threshold, this situation is not commonly encountered in practice, since the threshold would be generally above the soil field capacity for most of the soils. Hence, this is not considered a major issue for the Rectangular sensor.

Table 4 summarises the correlations obtained for the Rectangular sensor and the dielectric probe against volumetric water content. A power fit between the capacitive reading and the volumetric water content describes the correlations and indicates a good agreement (refer to Figure 4).

Table 4. Summary of the Rectangular sensor and dielectric probe performance against each soil sample.

Soil Sample	Rectangular Sensor			Dielectric Probe		
	Equation	R^2	RMSE (%)	Equation	R^2	RMSE (%)
Brighton Group Sand	$\theta = 0.63C^{0.77}$	0.96	4.11	$\theta = 0.057C^2 + 2.48C - 3.92$	0.98	1.08
Silty Sand	$\theta = 1.03C^{0.65}$	0.96	3.52	$\theta = -0.025C^2 + 2.04C - 2.5$	0.93	2.86
Silty Clay (Camb)	$\theta = 1.44C^{0.65}$	0.92	4.95	$\theta = 0.015C^2 + 2.09C - 2.5$	0.95	4.17
Clayey Silt (Bun)	$\theta = 2.22C^{0.60}$	0.78	11.22	$\theta = 0.034C^2 + 2.34C - 2.34$	0.98	1.80

θ = Volumetric Water Content, C = Capacitance (pF), R^2 = Coefficient of Correlation, RMSE = Root Mean Square Error.

Based on these results, the Rectangular sensor appears to have effectively predicted the variation of the volumetric water content. However, the size of the sensor may hinder good soil-sensor coupling (e.g., Bun Silt). This is thought to be due to the relatively larger size of the Rectangular sensor which made working with samples such as Bun Silt harder where the sample surface presented large undulations.

The results of the cross-validation study for the Rectangular sensor are summarised in Table 5.

Table 5. Cross validation analysis for the Rectangular sensor for each soil.

Soil Sample	Calibration		Validation	
	R^2	RMSE (%)	R^2	RMSE (%)
Brighton Group Sand	0.96	3.40	0.97	5.35
Silty Sand	0.95	3.20	0.95	3.95
Silty Clay (Camb)	0.93	4.90	0.93	5.17
Clayey Silt (Bun)	0.78	10.32	0.78	14.18

The calibration function for each soil, which was based on half of the experimental data, is shown to be able to predict the behaviour of the remainder of the dataset with a strong correlation. Moreover, the RMSEs of the validation dataset are comparable to the ones from the calibration. This analysis further demonstrates the capability of the Rectangular sensor in predicting the variation in soil moisture content of different soil types. Nonetheless, there are limitations with regard to the multiple electrode design as well as the size of the sensor; these limitations were addressed in the development of the fourth sensor.

5.3. PCB Sensor

The PCB sensor had a large geometry and sensing area compared to the Circular sensor, however, was smaller than the Rectangular sensor. Therefore, maintaining good coupling between the soil and the PCB sensor was relatively easy due to its size and being a single electrode. Figure 5 depicts the variation of PCB sensor outputs with water content, alongside the measured dielectric constants for all of the samples. A clear trend between capacitive readings and volumetric water content is shown across the samples; however, with a much lower sensitivity to water content beyond 15%. For the soil samples shown in Figure 5, the change in sensor readings is significantly larger for a water content variation from dry to approximately 15%, than for a water content variation beyond 15%. This clearly indicates that the PCB sensor is able to distinguish the changes in the water content; however, with a significantly reduced ability to accurately estimate the water content beyond 15%. This further highlights the limited capability of this sensor to estimate the water content of the fine-grained soils tested here, which can generally have water content above 15% in natural conditions.

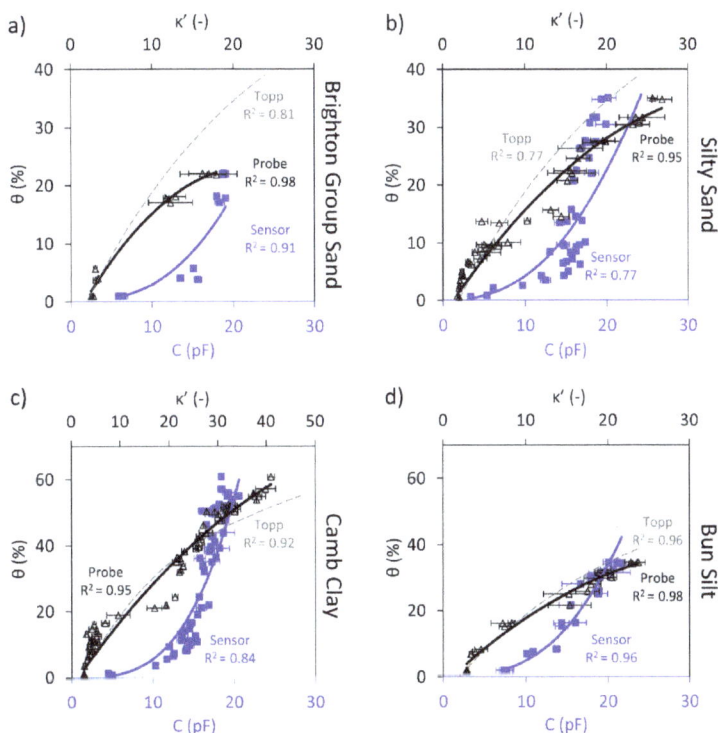

Figure 5. PCB sensor capacitance, C, readings (blue square markers and trendline) and dielectric constant, κ', measurements (black triangular markers and trendline) shown against the volumetric water content, θ; the Topp equation is also shown (dashed grey trendline). Shown for: (**a**) Brighton Group Sand (**b**) Silty Sand (**c**) Camb Clay and (**d**) Bun Silt.

Table 6 summarises the calibration obtained for the PCB sensor as well as for the dielectric probe. The values show that the performance of the PCB sensor in predicting the water content is comparable to that of the dielectric probe; nonetheless, with the aforementioned limitation regarding the measurements for samples with water content beyond 15%, indicated by higher errors despite high R^2 values (see Table 6).

Table 6. Summary of the PCB sensor and dielectric probe performance against each soil sample.

Soil Sample	PCB Sensor			Dielectric Probe		
	Equation	R^2	RMSE (%)	Equation	R^2	RMSE (%)
Brighton Group Sand	$\theta = 0.007C^{2.63C}$	0.91	3.96	$\theta = 0.057C^2 + 2.48C - 3.92$	0.98	1.08
Silty Sand	$\theta = 0.017C^{2.40}$	0.77	7.10	$\theta = -0.029C^2 + 2.11C - 2.5$	0.95	2.38
Silty Clay (Camb)	$\theta = 0.003C^{3.28}$	0.84	9.67	$\theta = 0.015C^2 + 2.09C - 2.5$	0.95	4.17
Clayey Silt (Bun)	$\theta = 0.008C^{2.8}$	0.96	3.75	$\theta = 0.034C^2 + 2.34C - 2.34$	0.98	1.80

θ = Volumetric Water Content, C = Capacitance (pF), R^2 = Coefficient of Correlation, RMSE = Root Mean Square Error.

Cross-validation analysis was performed for the PCB sensor and the results are summarised in Table 7. The results suggest that the PCB sensor is able to capture the increasing trend between the capacitive reading and volumetric water; however, the errors are relatively large. The large errors are assumed to be the result of the reduced sensitivity of the PCB sensor.

Table 7. Cross validation analysis for the PCB Sensor for each soil.

Soil Sample	Calibration		Validation	
	R^2	RMSE (%)	R^2	RMSE (%)
Brighton Group Sand	0.91	3.36	0.91	4.39
Silty Sand	0.76	6.72	0.77	7.10
Silty Clay (Camb)	0.855	9.44	0.86	9.86
Clayey Silt (Bun)	0.96	0.93	0.96	4.09

From these results, it is concluded that the geometry of the PCB sensor rectifies the sensor-soil contact issue. Moreover, the design is further enhanced by being a single electrode (similar to the Circular sensor) as opposed to being a multiple electrode design (e.g., the Rectangular sensor). However, there is the issue of sensitivity, caused by the controller board of the PCB sensor, which precludes accurate and reliable estimation for soils with water content above 15%.

5.4. AOGAN Sensor—An Integrated Sensor Designed Utilising the Advantages of the Previous Sensors

It is shown in the previous sections that the Circular, Rectangular and PCB sensors are able to capture changes in soil water content. However, there are advantages and limitations associated with each of the sensors.

In summary, the advantages are as follows: Firstly, a single electrode helped to maintain superior soil-sensor contact (Circular and PCB sensors) and showed a lower sensitivity to high frequency surface undulations relative to the sensor size (PCB sensor). Secondly, the controller board of the Rectangular sensor provided a reliable capacitive sensing range, facilitated stable readings and enabled working with samples with high water content. Moreover, the insulating agent coating the Rectangular sensor effectively prevented potential short circuit issues when dealing with very wet soil samples. Similar to that of the PCB sensor, the larger geometry of the Rectangular sensor created a better platform for conducting the measurements and maintaining good soil-sensor coupling or contact. Nonetheless, the significantly larger design of the Rectangular sensor proved to be problematic for some measurements (e.g., Bun Silt samples). Furthermore, from the perspective of conducting the measurements, the shape of the Circular sensor proved to be superior than that of the other two sensors, in maintaining contact and in the ease of use.

On the other hand, key limitations include the inadequate sensing range and the instability issue of the Circular sensor, the multiple electrodes and significantly larger geometry of the Rectangular sensor and the limited sensitivity issue of the PCB sensor.

The AOGAN sensor was designed by combining the identified advantages of the Circular, Rectangular and PCB sensors and eliminating their identified limitations. As such, the design of the AOGAN sensor was inspired by the shape and single electrode design of the Circular sensor and the larger geometry of the PCB sensor and incorporated a board designed and printed to act as the sensing component. This sensing component was almost three times larger than that of the Circular sensor to help with increasing the sensing range and was accompanied by a waterproof agent (similar to the Rectangular and the PCB sensors) to eliminate the potential short circuit issue. It is important to note that using an insulating film creates a sensor that is measuring two capacitors in series: formed by the insulating film and the soil, respectively. However, since the thickness of the insulating film was less than 0.05 mm, we assumed that the effect on the soil water content estimation was minimal. The larger geometry and the single electrode design improved practicality for conducting measurements. Furthermore, the sensing component was controlled by a board similar to the one used in the Rectangular sensor (which provided a more stable and larger sensing range and readings). The sensing component and the controller board were then connected to an Arduino-based board which communicated with a laptop. The program used for controlling the AOGAN sensor was the same as the one used for the Rectangular sensor.

An experiment was designed to estimate the sensing range of the sensor. Based on the methodology described by Orangi and Narsilio [71], a wet soil sample was prepared with the Silty Sand and was placed on a lab jack. The initial distance between the sample and the sensor was at 50 mm and the sample has subsequently approached the sensor at small increments controlled by a dial gauge. The capacitance measurements were recorded until a full contact between the soil and sensor was achieved. The result of this experiment is shown in Figure 6. The normalised sensor output, S_N, is plotted against the separation between the soil sample and the sensor, Δ. The figure shows that the sensing range is within 10 to 16 mm. We have estimated, therefore, that the depth that the sensor is able to estimate soil water content is around 10 mm. It is worth mentioning that the full contact between the soil and the sensor resulted in a more reliable sensor output.

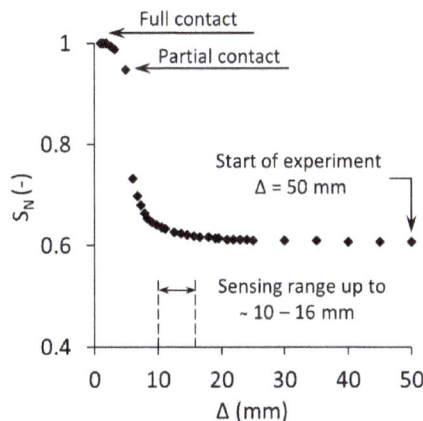

Figure 6. Normalised Sensor reading, S_N, against the distance between the soil sample and sensor Δ.

Figure 7 depicts the strong correlations obtained between the AOGAN sensor capacitive readings and water contents for all of the samples that were tested. Minimal issues were encountered with regard to soil-sensor coupling, instability, sensing ranges and low sensitivity to water content variation, which were observed for the previous sensors. It is assumed that this enhanced performance is a result of the advantageous features incorporated in the design of the AOGAN sensor.

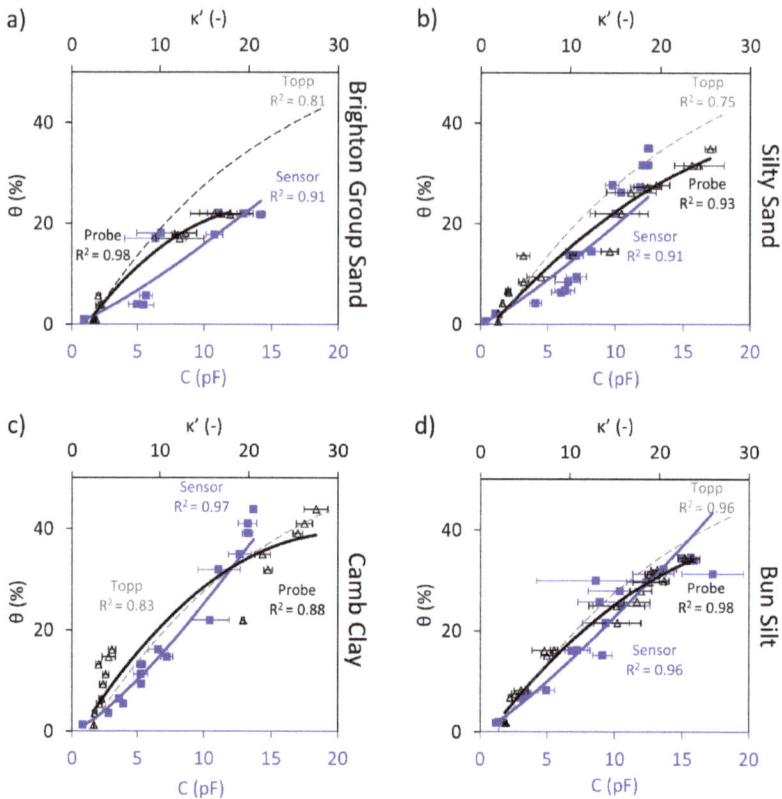

Figure 7. AOGAN sensor capacitance, C, readings (blue square markers and trendline) and dielectric constant, κ', measurements (black triangular markers and trendline) shown against the volumetric water content, θ; the Topp equation is also shown (dashed grey trendline). Shown for: (**a**) Brighton Group Sand (**b**) Silty Sand (**c**) Camb Clay and (**d**) Bun Silt.

It is important to note that the preparation of samples plays a key role in for the surface quality status of the soil samples, which impacts the quality of the soil-sensor contact or coupling. In practical applications, it is crucial to note that the deployment of sensors in the field requires the development of a mechanism that maintains full soil-sensor contact. This is to ensure a reliable sensor performance, as was observed during the laboratory measurements where the contact was maintained by using a weight.

As shown in Figure 7d, there are some samples for which the standard deviations of the measurements (error bars) are relatively large, possibly due to the contact issue between the soil and the sensor. These are some of the same samples for which the rectangular sensor was unable to capture the soil water content variations.; however, the AOGAN sensor showed less sensitivity to surface undulations due to its smaller size. The large standard deviations may, therefore, be the result of inadequate trimming of the samples and not the sensor hardware. Additionally, the inconsistency in the soil water content of a given sample was most likely not the reason for such discrepancies, since the samples were mixed thoroughly and cured during the preparation stage. As a result, it can be assumed that the water content was relatively homogenous for a given sample. The errors involved in soil water content estimation based on the calibration for this sensor were less than 5%. A summary of the results is shown in Table 8.

Table 8. Summary of the AOGAN sensor and dielectric probe performance against each soil sample.

Soil Sample	AOGAN Sensor			Dielectric Probe		
	Equation	R^2	RMSE (%)	Equation	R^2	RMSE (%)
Brighton Group Sand	$\theta = 0.94C^{1.23}$	0.91	3.88	$\theta = 0.057\kappa'^2 + 2.48C - 3.92$	0.98	1.08
Silty Sand	$\theta = 1.41C^{1.15}$	0.91	4.8	$\theta = -0.025C^2 + 2.04C - 2.5$	0.93	2.86
Silty Clay (Camb)	$\theta = 1.23C^{1.31}$	0.97	2.78	$\theta = 0.045C^2 + 2.74C - 2.5$	0.88	4.96
Clayey Silt (Bun)	$\theta = 1.49C^{1.18}$	0.96	4.4	$\theta = 0.034\ C^2 + 2.34C - 2.34$	0.98	1.80

θ = Volumetric Water Content, C = Capacitance (pF), R^2 = Coefficient of Correlation, RMSE = Root Mean Square Error.

Overall, considering the improved performance attributes, the AOGAN sensor shows great potential to estimate the soil water content non-invasively.

The cross-validation analysis results for the AOGAN sensor are given in Table 9. Overall, the statistical measures for the calibration and validation functions show the strong capability of the AOGAN sensor to capture the variations of the volumetric water content of soils. The values in the table also show the superior performance of this sensor compared to the other three sensors (See Tables 3, 5 and 7).

Table 9. Cross validation analysis for the AOGAN sensor for each soil.

Soil Sample	Calibration		Validation	
	R^2	RMSE (%)	R^2	RMSE (%)
Brighton Group Sand	0.89	3.83	0.89	4.66
Silty Sand	0.92	4.13	0.92	4.79
Silty Clay (Camb)	0.97	2.48	0.97	3.07
Clayey Silt (Bun)	0.96	4.02	0.96	4.75

5.5. Effect of Soil Type on the Calibration of the Sensor

It is reported in the literature that the relationship between a soil's electrical properties and its water content is determined by the soil type [81–83]. It is, therefore, imperative to evaluate the extent of soil type effects on the performance of the sensors in this work. In the results section, we showed that for sensors with good performance, a separate calibration could adequately describe the data for each soil sample. However, employing a single calibration for each individual soil type may not be practical and could become a tedious task in practice. This likely explains why a single calibration has been adopted for a range of soil types in previous studies (e.g., see [53,54,84,85]). The empirical Topp calibration proposed by Topp, Davis and Annan [53] was based on soils ranging from heavy clay to sandy loam; however, it is unable to accurately estimate the water content of some soil samples tested in the current study and also in a number of previous studies (e.g., [86–90]). Nonetheless, it is currently one of the most widely used empirical calibrations for estimating water content using soil dielectric properties. Therefore, following the same approach, we evaluated the efficacy of using a single calibration for each sensor in the present study. Data from the four samples was collated as a dataset and the performance of each sensor and the dielectric probe was analysed against it.

Figure 8 shows the collated capacitance readings and real dielectric constant data for the four samples plotted against volumetric water content. The Topp calibration is also shown for comparison. For clarity, the standard deviations of the measurements have been removed. The collation of capacitive data is shown by highlighted yellow markers and different markers correspond to different soil types, captured by a blue trendline. The capacitive readings are fitted with a power function as in the previous section. The measured dielectric constants and the corresponding trendlines are shown by black triangular markers and black trendlines, respectively.

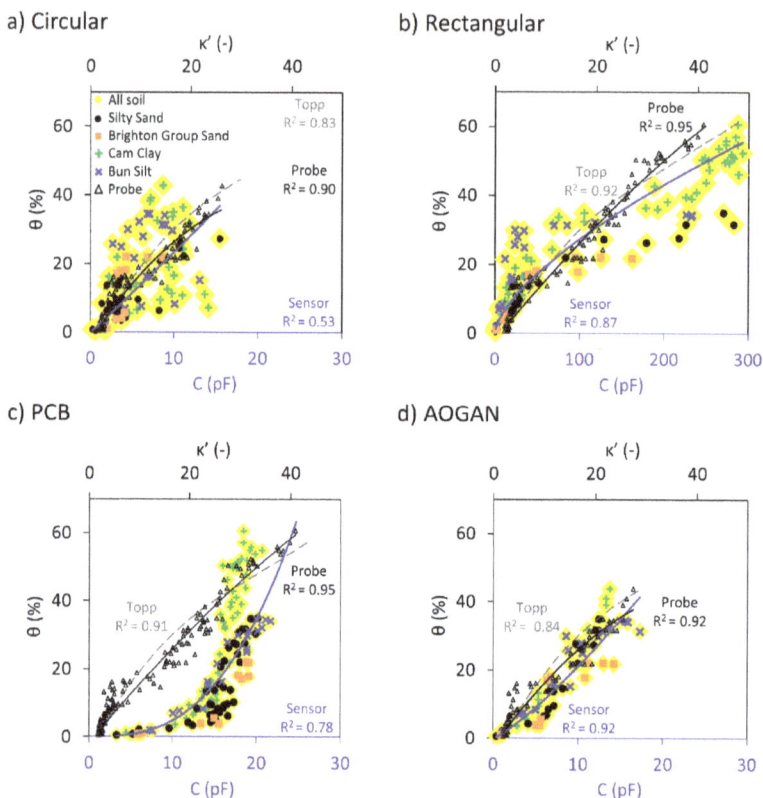

Figure 8. Combined capacitance, C and dielectric constant, κ′, data versus volumetric water content, θ, for: (**a**) Circular, (**b**) Rectangular, (**c**) PCB and (**d**) AOGAN sensors. Highlighted yellow markers capture the capacitive data for all soils and different markers represent different soil samples captured by the blue trendline. Black triangular markers and dashed trendlines correspond to dielectric measurements. Topp calibration is shown by the dashed grey trendlines.

As seen previously, the Circular sensor was unable to capture the variation of soil water content for the cohesive samples (Figure 8a); it was better able to capture the trend for the sandy samples. In addition, the instability issue occurred for all of the soil samples. Therefore, due to these two issues, combining the data for this sensor to evaluate the efficacy of a single calibration has resulted in a weak correlation ($R^2 = 0.53$) and large errors ($\approx$10%) as presented in Table 10.

Table 10. The response of each sensor against the combined dataset. The response of the dielectric constant data against the corresponding dataset used for each sensor is also shown.

	Sensor			Dielectric Probe		
Sensor	**Equation**	**R^2**	**RMSE (%)**	**Equation**	**R^2**	**RMSE (%)**
Circular	$\theta = 2.09C^{1.05}$	0.53	9.92	$\theta = 0.029\kappa'^2 + 2.21\kappa' - 2.5$	0.90	3.51
Rectangular	$\theta = 1.31C^{0.66}$	0.87	7.35	$\theta = -0.008\kappa'^2 + 1.82\kappa' - 2$	0.95	3.70
PCB	$\theta = 0.006C^{2.78}$	0.78	11.94	$\theta = -0.007\kappa'^2 + 1.79\kappa' - 2$	0.95	3.6
AOGAN	$\theta = 1.24C^{1.23}$	0.92	4.85	$\theta = -0.023\kappa'^2 + 2.09\kappa' - 2.5$	0.92	3.54

θ = Volumetric Water Content, C = Capacitance (pF), R^2 = Coefficient of Correlation, RMSE = Root Mean Square Error.

By contrast, for the Rectangular sensor shown in Figure 8b, a single calibration describes the dataset with relatively good agreement and resembles the calibration obtained using the dielectric constant data. Nonetheless, using the single calibration for the sensor may clearly lead to overestimating the water content of the sand samples (i.e., Brighton Group Sand and Silty Sand samples that are shown by orange and black markers, respectively) for soils with moisture content above approximately 20%. This is highlighted in Table 10 showing a lower R^2 value ($R^2 = 0.87$) and relatively large errors ($\approx$7.5%) compared to the calibration for individual soil samples (Table 4). Considering the capacitive measurements obtained by this sensor, one can see two distinct clusters of data versus volumetric water content. These two clusters, in fact, can be categorised as sandy soils and cohesive soils. Thus, two separate calibrations are expected to better estimate the water content for each cluster.

With regard to the PCB sensor, it can be seen in Table 10 that employing a single calibration for describing the capacitive reading versus volumetric water content has resulted in a lower R^2 than for the Rectangular sensor and larger errors. This further suggests that deriving separate calibrations for sand and clay groups can help in better describing the water content variation using the capacitive reading. However, the performance of the PCB sensor was proved to be questionable in the previous sections.

For the AOGAN sensor depicted in Figure 8d, a single calibration is shown to effectively capture the capacitive readings and volumetric water content relationship for the combined dataset (see Table 10). Compared to the previous sensors, due to a weaker contrast between the measurements made for the sand and clay groups, it is suggested that there is a lesser dependency on the soil type for the AOGAN sensor. The improved geometry and controller board design of the AOGAN sensor appears to provide better soil-sensor coupling and readings, which are, in turn, less affected by the soil type. Nonetheless, the data is treated separately in two groups in order to refine the correlations.

Table 10 includes a summary of the data corresponding to the sensors and the dielectric probe calibrations. These calibrations were derived and assessed against the corresponding dataset used for each sensor. It is worth mentioning that the relationship between volumetric water content and dielectric constant data could be adequately described by adopting single calibrations, which are superior to the Universal Topp calibration, as these here become site specific calibrations.

It is shown that the performance of the AOGAN sensor against the combined dataset is superior to that of the other sensors and that it is less affected by the soil type. However, it is suggested that adopting separate calibrations could provide improved predictions for the sensor. Moreover, for the other sensors, distinct behaviour was observed for the sand and clay groups and relatively larger errors were introduced by adopting a single calibration.

Figure 9 shows for each sensor the combined capacitive readings for all of the soils but data are grouped into sands (described by the black solid trendline) and clays (described by the blue dashed trendlines).

It was concluded from previous analyses in Sections 5.1 and 5.5 that the Circular sensor is unable to detect changes in water content of the clayey samples and therefore that the performance of this sensor is not satisfactory at least for the clayey samples. It is unsurprising, therefore, that for this sensor, better performance is observed when using separate calibrations. For the Rectangular sensor, separate calibrations seem to capture the individual groups, rendering smaller errors for each group and higher R^2 values for the sand group, as shown in Figure 9b and Table 11. For the PCB sensor, the predictive performance has smaller errors for both the sand and clay groups compared with the single calibration; whereas, the performance of the AOGAN sensor is less dependent on the soil type, as shown by only a marginal improvement upon employing separate calibrations.

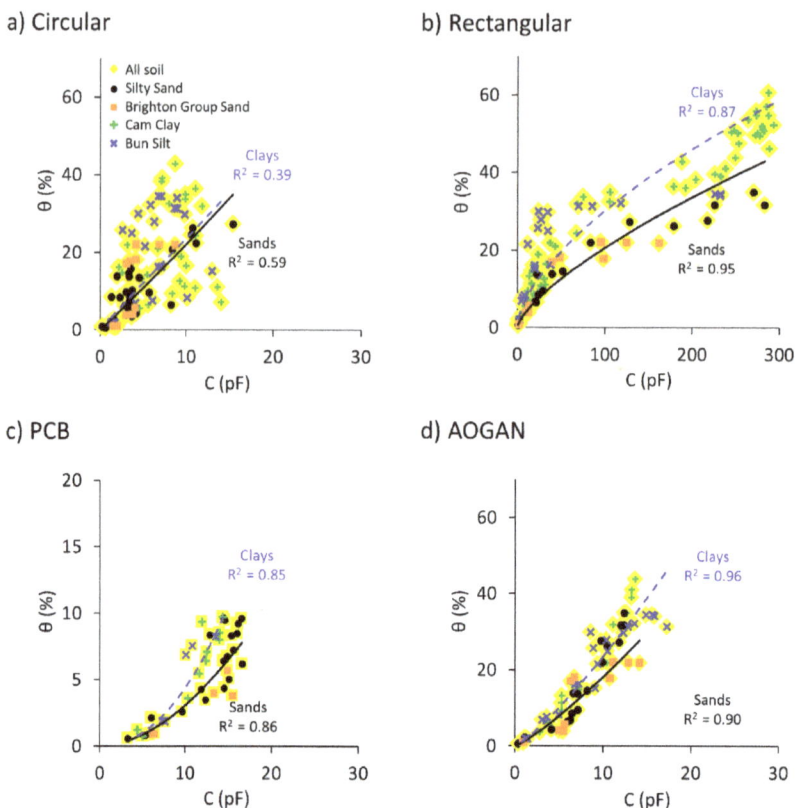

Figure 9. Combined capacitance, C and dielectric constant, κ', data versus volumetric water content, θ, for the: (**a**) Circular, (**b**) Rectangular, (**c**) PCB and (**d**) AOGAN sensors. Highlighted yellow markers capture the capacitive data for all soils and different markers represent different soil samples. Separate calibrations are used for describing the sand group (black solid trendline) and clay group (blue dashed trendline).

Table 11. Performance of the sensors against combined soil, combined sand group and combined clay group data.

Sensor	All Soils			Sands			Clays		
	Equation	R^2	RMSE (%)	Equation	R^2	RMSE (%)	Equation	R^2	RMSE (%)
Circular	$\theta = 2.09C^{1.05}$	0.53	9.92	$\theta = 1.99C^{1.04}$	0.59	5.34	$\theta = 2.43C^{0.99}$	0.39	12.36
Rectangular	$\theta = 1.31C^{0.66}$	0.87	7.35	$\theta = 0.81C^{0.71}$	0.95	4.20	$\theta = 1.82C^{0.61}$	0.87	6.78
PCB	$\theta = 0.006C^{2.78}$	0.78	11.94	$\theta = 0.01C^{2.51}$	0.81	6.61	$\theta = 0.004C^{3.11}$	0.85	11.39
AOGAN	$\theta = 1.24C^{1.23}$	0.92	4.85	$\theta = 1.14C^{1.21}$	0.90	4.95	$\theta = 1.39C^{1.23}$	0.96	4.62

It is shown that by employing separate calibrations, all of the sensors obtain better goodness-of-fit and lower error in soil water content estimation. Nonetheless, the predictive performance of the AOGAN sensor improved the least by adopting this approach, which again highlights the superiority of the AOGAN sensor among the other sensors.

5.6. Dielectric Constant Approximation through Capacitive Measurements

It was shown in the previous sections that the readings from the four capacitive sensors correlate (with different goodness-of-fits; 0.57 for the Circular sensor, 0.91 for the Rectangular sensor, 0.87 for the

PCB sensor and 0.94 for the AOGAN sensor) with variation in the volumetric water content. On the other hand, it is known that the capacitance of a capacitor is derived by its geometry and the dielectric constant of its material (refer to Equation (4)). Therefore, can the dielectric constant of the material under test be reliably estimated using the capacitive sensor readings? In this section, correlations between the sensor readings and the dielectric constant data (measured by the Agilent dielectric slim form probe) are obtained, for both the sand and clay groups, for each sensor.

Figure 10 shows the sensor readings versus the dielectric constant data. For each sensor, the data are divided into sand and clay groups and for each a linear correlation which passes through the origin illustrates the relationship shown in Equation (4). This approach enables estimation of the geometry factor, g, for the sensors and the dielectric constant from the sensor output.

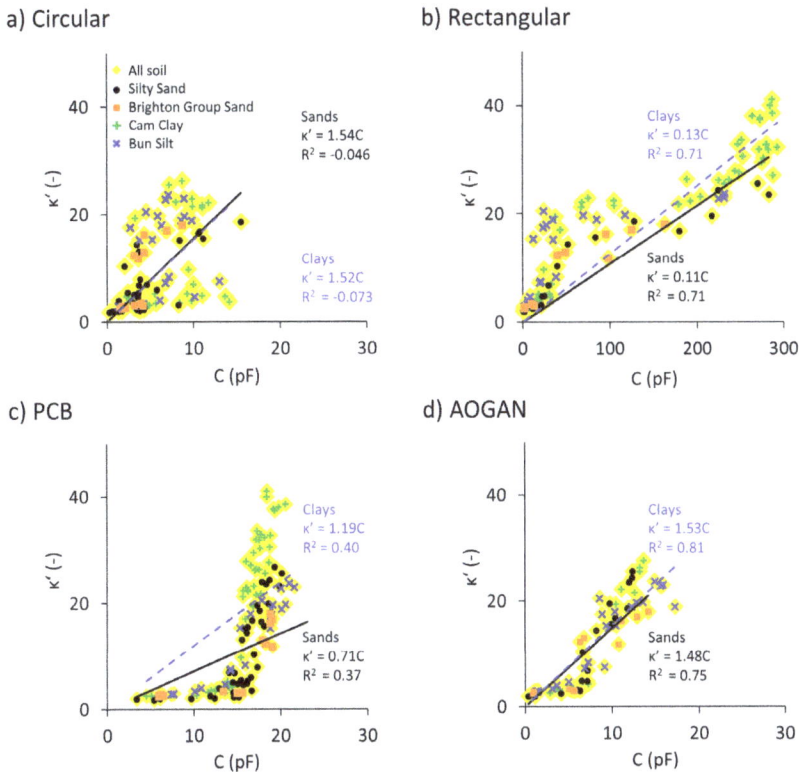

Figure 10. Measured dielectric constant, κ', data against capacitive, C, readings for the sand (black solid trendline) and clay (blue dashed trendline) groups, for the: (**a**) Circular, (**b**) Rectangular, (**c**) PCB and (**d**) AOGAN sensors.

The performance of the Circular sensor was shown to be unsatisfactory in the previous section and this is verified by the relationship obtained here, represented by the trendline. The Rectangular sensor shows stronger correlations for both soil groups. It is shown that the equations describing the relationships for the sand and clay groups are quite similar, which, in turn, result in similar geometry factors. The geometry factor for any capacitor is a constant value and this is somewhat explained by the similar slopes of the two trendlines of two soil groups. For the PCB sensor, as was explained previously, due to the weak signal-to-noise ratio (i.e., lower sensitivity), the correlation between the sensor reading and the measured dielectric constant is not strong for any of the soil groups. With regard to the AOGAN sensor, due to its improved design and performance, correlations between the sensor readings and the measured dielectric constant are strongest. In addition, the correlation between capacitive readings obtained from the AOGAN sensor and dielectric constant data are similar for both soil groups and return almost identical geometry factors (See Figure 10d).

The superior performance of the AOGAN sensor (followed by the Rectangular sensor) is demonstrated in this analysis by the strongest proportionality between the sensor readings and the measured dielectric constants.

In the next section, the effect of sample preparation on the performance of the sensors is discussed.

5.7. Effect of Surface Contact and Roughness

Contact between soil and a sensor and the roughness of a sample's surface, are demonstrated by the experimental data to be important factors influencing the quality of measurements made with the sensors. It was suggested by Orangi, Withers, Langley and Narsilio [80] that the soil-sensor coupling and soil surface roughness could be responsible for the poor performance of the Rectangular and PCB sensors in their work. Hence, in this section, the raw data from Orangi, Withers, Langley and Narsilio [80] are re-examined to investigate the importance of these factors.

In this present work, of the twelve readings obtained from a single measurement with the Rectangular sensor, the maximum reading was taken as the representative capacitance value instead of using the mean value. This decision was based on the result of the following analysis conducted on the experimental data published by Orangi, Withers, Langley and Narsilio [80] on the Brighton Group Sand. That is, for a set of experiments conducted on the Brighton Group Sand, we have adopted two methods for the analysis. Firstly, for each sample, the variation of capacitance with water content was investigated, using the mean value of the readings from 12 electrodes. The second approach was to use the maximum of the 12 readings obtained by the Rectangular sensor. The results are shown in Figure 11.

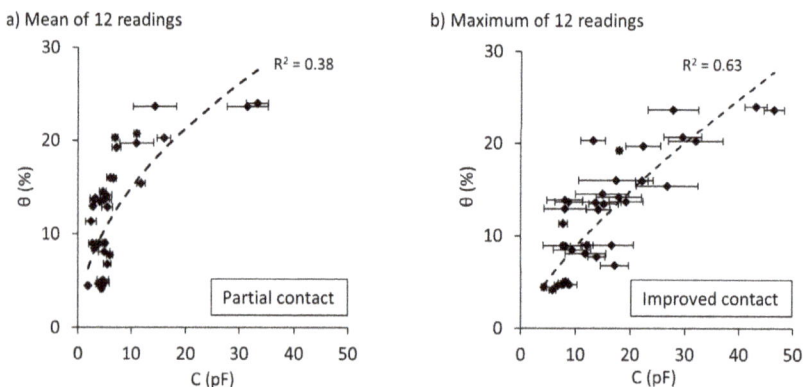

Figure 11. Effects of soil sensor coupling on the estimation of volumetric water content, θ, using the Rectangular sensor on the Brighton Group Sand: (**a**) Mean of 12 readings (Partial contact), (**b**) Maximum of 12 readings (Improved contact). Data adapted from [80].

As illustrated in Figure 11, the ability of the Rectangular sensor to detect water content variation is improved when the maximum value of the 12 readings is considered as the capacitance value (Figure 11b). It is observed that the R^2 value increases by more than 35% when choosing the maximum sensor reading instead of the mean (Figure 11a). Nonetheless, for the Rectangular sensor and the Brighton Group Sand, the value of R^2 obtained in the above study ($R^2 = 0.63$) is less than the value obtained in the current study ($R^2 = 0.95$, Figure 4a, Table 4). This is likely attributed to partial contacts exist during the previous experimentation. Therefore, these results suggest that soil-sensor coupling can significantly affect the performance of the sensors. The issue associated with the number of electrodes discussed in Section 4.2 does not apply to the PCB sensor or other sensors comprising a single electrode; however, by using the data for the Rectangular sensor we can demonstrate the importance of effective coupling, as well as justifying the rationale behind utilising the maximum output as the reading of the Rectangular sensor.

The surface of the soil samples tested by Orangi, Withers, Langley and Narsilio [80] were less smooth than those used in the current study. Figure 12 shows the relationship between the Rectangular sensor readings and the water content data for a Basaltic Clay tested by Orangi, Withers, Langley and Narsilio [80] and for the Camberwell Clay tested in this work. Figure 12a shows the capacitive values (recorded as the *mean* of the 12 readings) of the Basaltic Clay samples tested by Orangi, Withers, Langley and Narsilio [80], which had *rough* surfaces. It can be seen in this figure that the correlation between the sensor readings and the water content is very weak. Figure 12b shows the *mean* capacitive readings for the Camb Clay, which had *smooth* surfaces. It can be seen that the correlation between the sensor readings and the water content improves significantly with a smoother soil surface. However, the soil-sensor(s) coupling is considered to not be evenly maintained, due to using the mean of 12 readings. Figure 12c shows data for the Basaltic Clay in which the *maximum* of the 12 readings was taken as the sensor reading. Improved contact and correlation can be seen compared with Figure 12a; however, since the surfaces of the samples were *rough* the correlation is still weak.

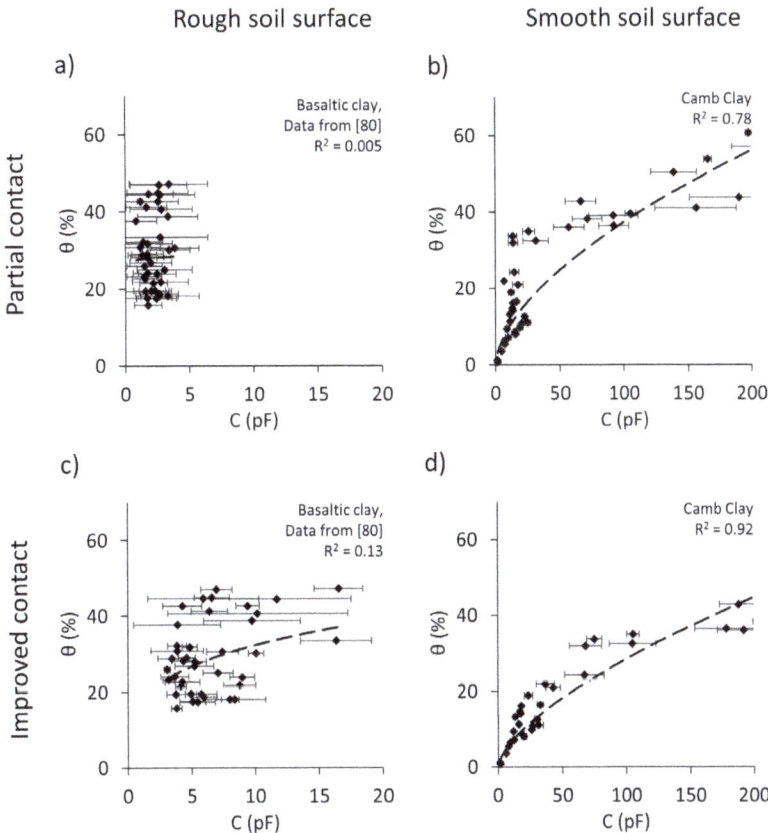

Figure 12. Effect of soil surface roughness and sensor contact on the performance of sensors for two clay samples: Basaltic Clay and Camb Clay samples. Basaltic Clay data are from Orangi, Withers, Langley and Narsilio [80].

The weak performance observed in Figure 12a,c is due to the rough soil surfaces associated with sample preparation. Moreover, the standard deviation of the capacitance values from multiple measurements made for a given sample are larger compared to the results of the current study. This appears to be the result of samples having uneven surfaces, creating large variations in sensors readings. However, in Figure 12d, a significant improvement in the quality of the data is shown, owing to the *smoothness* of the sample surfaces as well as good soil-sensor contact, achieved by taking the *maximum* measured value of the 12 readings. Therefore, by comparing the results of the current study with the previous study [80] for two clay samples using the Rectangular sensors, it can be concluded that the surface roughness of samples and the soil-sensor contact play vital roles in ensuring reliable measurements.

The Camb Clay and Bun Silt samples had similar surface finishes. Moreover, Figure 13d shows an example of the surfaces of clay samples (i.e., Basaltic Clay) used by Orangi, Withers, Langley and Narsilio [80]. The surface roughness visible in this figure was a result of inadequate trimming of the samples during the preparation step by Orangi, Withers, Langley and Narsilio [80]. It is thought to be the underlying reason for the weak performance of the Rectangular and PCB sensors shown in Figure 12, compared to the results of the current study.

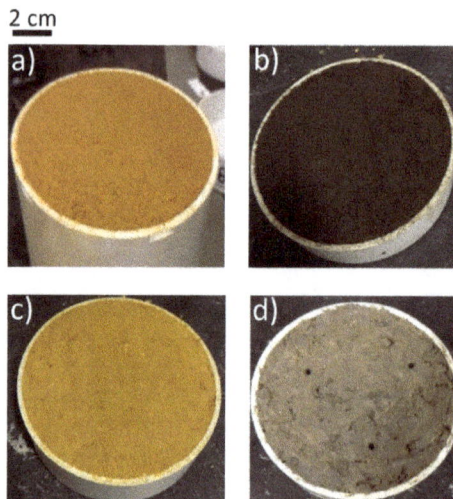

Figure 13. Typical examples of the prepared samples of: (**a**) Brighton Group Sand, (**b**) Silty Sand and (**c**) Camb Clay from the present study and (**d**) Basaltic Clay from Orangi, Withers, Langley and Narsilio [80].

Based on the analysis conducted in this section, it is suggested that a smooth surface as well as full coupling between the soil sample surface and the sensor are key factors in ensuring that sensors can effectively estimate the soil water content and detect its variations.

6. Potential Applications and Limitations

The new sensors developed in this work do not require insertion into the soil, nor do they require subsamples to be retrieved for subsequent gravimetric calculations. Furthermore, due to their non-invasive nature and the speed of measurements (milliseconds to retrieve a reading), repetitive measurements at the exact same location are possible, which makes these sensors suitable for large-scale near surface soil water content monitoring. A potential application in the agriculture sector could be high spatial and temporal resolution mapping of surface soil moisture, beneficial for farm management. The non-invasive characteristic of the sensors enables frequent soil moisture measurement across a farm, which aids in decision making concerning sowing time, irrigation and fertiliser scheduling. Moreover, the sensors can provide ground truth data to calibrate satellite image and remotely sensed soil moisture data. Additionally, the new sensors can be used as a real-time monitoring system of near surface soil water content, for quantifying the risk of bushfire and generating warnings to the pertinent authorities when the soil water content falls below a certain threshold. Another potential application includes the use of the new sensors as a quick way of estimating the moisture content of sub-base and stockpile materials in the road construction industry. However, regardless of the application, the quality of the surface where the measurement is conducted against is an imperative parameter. Therefore, a smooth surface must be somehow achieved in the agricultural fields. With the current sensor design, however, it may be impractical to be used as a field sensor. Thus, it is deemed necessary that further mechanical features are added to the sensor to help with soil surface preparation in the field. On the other hand, in road construction applications, when quality assurance (i.e., water content and density measurements) is conducted, the surface of the sub-base layer after each run is considered to be adequately flat and smooth for direct measurement with this sensor.

Regarding limitations of the sensors, it is important to state that the estimation of soil water content is currently limited to approximately 1 cm deep, based on the width of the electrodes, W and their spacing, S, estimated using the approach proposed by Gao, Zhu, Liu, Qian, Cao and Ni [49]. Thus, measurements of deep soil moisture (e.g., up to approximately 1 m), which may be required in precision agriculture, particularly for the horticulture sector, cannot be conducted directly by the sensor from the surface. However, there are crops with shallower root zones, such as vegetables, whose management could benefit from shallower soil water content information. Furthermore, in terms of the accuracy of the data, the errors involved in the soil water content estimation using the AOGAN sensor were found to be in the order of 1 to 5%. Despite the sensor not being able to satisfy the desirable 1% resolution of soil water content data for precision agriculture, suggested by Terry A. Brase [91], it can still be used as a mapping tool for providing comparative assessments of soil water content on large scales. Moreover, accessing surface soil moisture information across large areas and frequently in time allows calibration of evapotranspiration soil models for continuous estimation of soil moisture with depth, valuable for agricultural and hydrological applications (e.g., see [19,27–29]).

Overall, whilst considering the aforementioned limitations, these sensors show promising potential in estimating surface soil water content, with implications in diverse fields including agriculture, bushfire protection management and road construction.

7. Conclusions

The non-invasive estimation of soil water content using capacitive-based sensors was investigated in this research. The experimental program entailed testing four new capacitive sensors, the AD7747 (Circular), MPR121 (Rectangular), PCB and AOGAN sensors, against four soil types. Measurements of the dielectric constant of the samples with an Agilent slim form dielectric probe connected to a FieldFox network analyser aided in the comparison of results, analysis and calibration of sensors. The AOGAN sensor was designed and manufactured based on the key advantages of each of the AD7747 (Circular), MPR121 (Rectangular) and PCB sensors. Promising capabilities were observed for the Rectangular and AOGAN sensors, with relatively small errors to estimate the soil water content, particularly for the latter sensor. The effect of soil type on the performance of the sensor was tested by combining data from the samples and it appeared that a single calibration could be adopted to estimate the soil water content. However, adopting a single calibration for all of the four samples resulted in inferior sensor performance compared to when adopting individual calibrations for sand and clay groups separately. Finally, it was demonstrated that the performance of each of the sensors was affected by the level of contact maintained between the sensor and the soil surface and more importantly by the roughness of the soil surface which impacted the soil-sensor contact area.

Sensors **2019**, *19*, 651

Author Contributions: A.O. was primarily responsible for the design, manufacturing, and assembly, of the sensors, undertaking the experimental program, data collection, subsequent data analysis and writing the manuscript. G.A.N. contributed to the design and manufacturing of the sensor, designing the experimental program, data analysis and writing and editing the manuscript. D.R.'s contribution was towards data analysis and editing the manuscript.

Funding: This research was partially funded by the Hong Kong Research Grants Council (project no. T22-603/15-N), the 2016 Endeavour Australia Cheung Kong Research Fellowship from the Department of Education and Training of the Australian Government and the Australian Government Research Training Program Scholarship.

Acknowledgments: The first author would like to acknowledge the financial support provided by the 2016 Endeavour Australia Cheung Kong Research Fellowship while studying at The Hong Kong University of Science and Technology and the financial support from the Australian Government Research Training Program Scholarship, while studying at The University of Melbourne. Professor Yu-Hsing Wang is thanked for his insightful comments during the final stages of this study. Graduate student R. Fullerton is thanked for his help with some of the experimental work. N. Ballis assisted with proofreading and editing.

Conflicts of Interest: The authors declare no conflicts of interest.

References

1. Famiglietti, J.S.; Ryu, D.; Berg, A.A.; Rodell, M.; Jackson, T.J. Field observations of soil moisture variability across scales. *Water Resour. Res.* **2008**, *44*. [CrossRef]
2. Yu, X.; Drnevich, V.P. Soil water content and dry density by time domain reflectometry. *J. Geotechn. Geoenviron. Eng.* **2004**, *130*, 922–934. [CrossRef]
3. Rajib, M.A.; Merwade, V.; Yu, Z. Multi-objective calibration of a hydrologic model using spatially distributed remotely sensed/in-situ soil moisture. *J. Hydrol.* **2016**, *536*, 192–207. [CrossRef]
4. Han, J.; Mao, K.; Xu, T.; Guo, J.; Zuo, Z.; Gao, C. A soil moisture estimation framework based on the cart algorithm and its application in china. *J. Hydrol.* **2018**, *563*, 65–75. [CrossRef]
5. Vereecken, H.; Huisman, J.A.; Bogena, H.; Vanderborght, J.; Vrugt, J.A.; Hopmans, J.W. On the value of soil moisture measurements in vadose zone hydrology: A review. *Water Resour. Res.* **2008**, *44*. [CrossRef]
6. Western, A.W.; Grayson, R.B.; Blöschl, G. Scaling of soil moisture: A hydrologic perspective. *Ann. Rev. Earth Planet. Sci.* **2002**, *30*, 149–180. [CrossRef]
7. Fischer, G.; Tubiello, F.N.; van Velthuizen, H.; Wiberg, D.A. Climate change impacts on irrigation water requirements: Effects of mitigation, 1990–2080. *Technol. Forecast. Soc. Chang.* **2007**, *74*, 1083–1107. [CrossRef]
8. Visconti, F.; de Paz, J.M.; Martínez, D.; Molina, M.J. Laboratory and field assessment of the capacitance sensors decagon 10hs and 5te for estimating the water content of irrigated soils. *Agric. Water Manag.* **2014**, *132*, 111–119. [CrossRef]
9. Hedley, C.B.; Yule, I.J. A method for spatial prediction of daily soil water status for precise irrigation scheduling. *Agric. Water Manag.* **2009**, *96*, 1737–1745. [CrossRef]
10. Muñoz-Carpena, R.; Shukla, S.; Morgan, K. *Field Devices for Monitoring Soil Water Content*; Institute of Food and Agricultural Sciences, University of Florida Cooperative Extension Service, EDIS: Gainesville, FL, USA, 2004; Volume 343.
11. Bryson, L.; Jean-Louis, M.; Gabriel, C. Determination of in situ moisture content in soils from a measure of dielectric constant. *Int. J. Geotech. Eng.* **2013**, *6*, 251–259. [CrossRef]
12. Kumar, V.; Dharssi, I. *Evaluation of Daily Soil Moisture Deficit Used in Australian Forest Fire Danger Rating System*; Bureau of Meteorology: Melbourne, Australia, 2017.
13. Chow, V.; Maidment, D.; Mays, L. *Applied Hydrology*, 1st ed.; McGraw-Hill Science/Engineering/Math: New York, NY, USA, 1988; p. 572.
14. Dobson, M.C.; Ulaby, F.T. Active microwave soil moisture research. *IEEE Trans. Geosci. Remote Sens.* **1986**, *GE-24*, 23–36. [CrossRef]
15. Brown, R.J.; Brisco, B.; Leconte, R.; Major, D.J.; Fischer, J.A.; Reichert, G.; Korporal, K.D.; Bullock, P.R.; Pokrant, H.; Culley, J. Potential applications of radarsat data to agriculture and hydrology. *Can. J. Remote Sens.* **1993**, *19*, 317–329. [CrossRef]
16. O'Brien, E. An agricultural application of regional scale soil moisture modelling and monitoring. In *Soil Moisture Modelling and Monitoring for Regional Planning*; National Hydrology Research Centre: Saskatoon, SK, Canada, 1992; pp. 13–20.

17. Hawley, M.E.; Jackson, T.J.; McCuen, R.H. Surface soil moisture variation on small agricultural watersheds. *J. Hydrol.* **1983**, *62*, 179–200. [CrossRef]

18. Gillies, R.R.; Kustas, W.P.; Humes, K.S. A verification of the 'triangle' method for obtaining surface soil water content and energy fluxes from remote measurements of the normalized difference vegetation index (NDVI) and surface e. *Int. J. Remote Sens.* **1997**, *18*, 3145–3166. [CrossRef]

19. Wigneron, J.-P.; Olioso, A.; Calvet, J.-C.; Bertuzzi, P. Estimating root zone soil moisture from surface soil moisture data and soil-vegetation-atmosphere transfer modeling. *Water Resour. Res.* **1999**, *35*, 3735–3745. [CrossRef]

20. Robinson, M.; Dean, T.J. Measurement of near surface soil water content using a capacitance probe. *Hydrol. Process.* **1993**, *7*, 77–86. [CrossRef]

21. Colliander, A.; Jackson, T.J.; Bindlish, R.; Chan, S.; Das, N.; Kim, S.B.; Cosh, M.H.; Dunbar, R.S.; Dang, L.; Pashaian, L.; et al. Validation of smap surface soil moisture products with core validation sites. *Remote Sens. Environ.* **2017**, *191*, 215–231. [CrossRef]

22. Frangi, J.-P.; Richard, D.-C.; Chavanne, X.; Bexi, I.; Sagnard, F.; Guilbert, V. New in situ techniques for the estimation of the dielectric properties and moisture content of soils. *C. R. Geosci.* **2009**, *341*, 831–845. [CrossRef]

23. Chanzy, A.; Bruckler, L. Significance of soil surface moisture with respect to daily bare soil evaporation. *Water Resour. Res.* **1993**, *29*, 1113–1125. [CrossRef]

24. Daamen, C.C.; Simmonds, L.P. Measurement of evaporation from bare soil and its estimation using surface resistance. *Water Resour. Res.* **1996**, *32*, 1393–1402. [CrossRef]

25. Ines, A.V.M.; Mohanty, B.P. Near-surface soil moisture assimilation for quantifying effective soil hydraulic properties using genetic algorithm: 1. Conceptual modeling. *Water Resour. Res.* **2008**, *44*. [CrossRef]

26. Thoma, D.P.; Moran, M.S.; Bryant, R.; Rahman, M.; Holifield-Collins, C.D.; Skirvin, S.; Sano, E.E.; Slocum, K. Comparison of four models to determine surface soil moisture from c-band radar imagery in a sparsely vegetated semiarid landscape. *Water Resour. Res.* **2006**, *42*. [CrossRef]

27. Noilhan, J.; Planton, S. A simple parameterization of land surface processes for meteorological models. *Mon. Weather Rev.* **1989**, *117*, 536–549. [CrossRef]

28. Jackson, T.J. Profile soil moisture from surface measurements. *J. Irrig. Drain. Division Am. Soc. Civ. Eng.* **1980**, *106*, 81–92.

29. Montaldo, N.; Albertson, J.D.; Mancini, M.; Kiely, G. Robust simulation of root zone soil moisture with assimilation of surface soil moisture data. *Water Resour. Res.* **2001**, *37*, 2889–2900. [CrossRef]

30. Entekhabi, D.; Nakamura, H.; Njoku, E.G. Solving the inverse problem for soil moisture and temperature profiles by sequential assimilation of multifrequency remotely sensed observations. *IEEE Trans. Geosci. Remote Sens.* **1994**, *32*, 438–448. [CrossRef]

31. Ryu, D.; Famiglietti, J.S. Characterization of footprint-scale surface soil moisture variability using gaussian and beta distribution functions during the southern great plains 1997 (SGP97) hydrology experiment. *Water Resour. Res.* **2005**, *41*. [CrossRef]

32. Walker, J.P.; Willgoose, G.R.; Kalma, J.D. One-dimensional soil moisture profile retrieval by assimilation of near-surface observations: A comparison of retrieval algorithms. *Adv. Water Resour.* **2001**, *24*, 631–650. [CrossRef]

33. Kodikara, J.; Rajeev, P.; Chan, D.; Gallage, C. Soil moisture monitoring at the field scale using neutron probe. *Can. Geotech. J.* **2013**, *51*, 332–345. [CrossRef]

34. Muñoz-Carpena, R.; Dukes, M.D. Automatic irrigation based on soil moisture for vegetable crops. In *Nutrient Management of Vegetable and Row Crops Handbook*; Department of Agricultural and Biological Engineering, UF/IFAS Extension: Gainesville, FL, USA, 2015; Volume 173.

35. Zhang, N.; Fan, G.; Lee, K.; Kluitenberg, G.; Loughin, T. Simultaneous measurement of soil water content and salinity using a frequency-response method. *Soil Sci. Soc. Am. J.* **2004**, *68*, 1515–1525. [CrossRef]

36. Robinson, D.; Campbell, C.; Hopmans, J.; Hornbuckle, B.; Jones, S.B.; Knight, R.; Ogden, F.; Selker, J.; Wendroth, O. Soil moisture measurement for ecological and hydrological watershed-scale observatories: A review. *Vadose Zone J.* **2008**, *7*, 358–389. [CrossRef]

37. Steelman, C.M.; Endres, A.L.; Jones, J.P. High-resolution ground-penetrating radar monitoring of soil moisture dynamics: Field results, interpretation, and comparison with unsaturated flow model. *Water Resour. Res.* **2012**, *48*, W09538. [CrossRef]

38. Grote, K.; Hubbard, S.; Rubin, Y. Field-scale estimation of volumetric water content using ground-penetrating radar ground wave techniques. *Water Resour. Res.* **2003**, *39*, 1321. [CrossRef]
39. Chen, R.P.; Chen, Y.M.; Xu, W.; Yu, X. Measurement of electrical conductivity of pore water in saturated sandy soils using time domain reflectometry (TDR) measurements. *Can. Geotech. J.* **2010**, *47*, 197–206. [CrossRef]
40. Wraith, J.M.; Robinson, D.A.; Jones, S.B.; Long, D.S. Spatially characterizing apparent electrical conductivity and water content of surface soils with time domain reflectometry. *Comput. Electron. Agric.* **2005**, *46*, 239–261. [CrossRef]
41. Huisman, J.; Hubbard, S.; Redman, J.; Annan, A. Measuring soil water content with ground penetrating radar: A review. *Vadose Zone J.* **2003**, *2*, 476–491. [CrossRef]
42. Wu, S.Y.; Zhou, Q.Y.; Wang, G.; Yang, L.; Ling, C.P. The relationship between electrical capacitance-based dielectric constant and soil water content. *Environ. Earth Sci.* **2011**, *62*, 999–1011. [CrossRef]
43. Dias, P.C.; Roque, W.; Ferreira, E.C.; Siqueira Dias, J.A. A high sensitivity single-probe heat pulse soil moisture sensor based on a single npn junction transistor. *Comput. Electron. Agric.* **2013**, *96*, 139–147. [CrossRef]
44. Matile, L.; Berger, R.; Wächter, D.; Krebs, R. Characterization of a new heat dissipation matric potential sensor. *Sensors* **2013**, *13*, 1137–1145. [CrossRef]
45. Dias, P.C.; Cadavid, D.; Ortega, S.; Ruiz, A.; França, M.B.M.; Morais, F.J.O.; Ferreira, E.C.; Cabot, A. Autonomous soil moisture sensor based on nanostructured thermosensitive resistors powered by an integrated thermoelectric generator. *Sens. Actuators A Phys.* **2016**, *239*, 1–7. [CrossRef]
46. Campbell, G.S.; Calissendorff, C.; Williams, J.H. Probe for measuring soil specific heat using a heat-pulse method. *Soil Sci. Soc. Am. J.* **1991**, *55*, 291–293. [CrossRef]
47. Bristow, K.L.; Campbell, G.S.; Calissendorff, K. Test of a heat-pulse probe for measuring changes in soil water content. *Soil Sci. Soc. Am. J.* **1993**, *57*, 930–934. [CrossRef]
48. Bristow, K.L.; Kluitenberg, G.J.; Horton, R. Measurement of soil thermal properties with a dual-probe heat-pulse technique. *Soil Sci. Soc. Am. J.* **1994**, *58*, 1288–1294. [CrossRef]
49. Gao, Z.; Zhu, Y.; Liu, C.; Qian, H.; Cao, W.; Ni, J. Design and test of a soil profile moisture sensor based on sensitive soil layers. *Sensors* **2018**, *18*, 1648. [CrossRef] [PubMed]
50. da Eduardo Ferreira, C.; de Oliveira, N.E.; Morais, F.J.O.; Carvalhaes-Dias, P.; Duarte, L.F.C.; Cabot, A.; Siqueira Dias, J.A. A self-powered and autonomous fringing field capacitive sensor integrated into a micro sprinkler spinner to measure soil water content. *Sensors* **2017**, *17*, 575.
51. França, M.B.d.M.; Morais, F.J.O.; Carvalhaes-Dias, P.; Duarte, L.C.; Dias, J.A.S. A multiprobe heat pulse sensor for soil moisture measurement based on pcb technology. *IEEE Trans. Instrum. Meas.* **2018**, *PP*, 1–8. [CrossRef]
52. Mancini, M.; Hoeben, R.; Troch, P.A. Multifrequency radar observations of bare surface soil moisture content: A laboratory experiment. *Water Resour. Res.* **1999**, *35*, 1827–1838. [CrossRef]
53. Topp, G.; Davis, J.; Annan, A.P. Electromagnetic determination of soil water content: Measurements in coaxial transmission lines. *Water Resour. Res.* **1980**, *16*, 574–582. [CrossRef]
54. Malicki, M.; Plagge, R.; Renger, M.; Walczak, R. Application of time-domain reflectometry (tdr) soil moisture miniprobe for the determination of unsaturated soil water characteristics from undisturbed soil cores. *Irrig. Sci.* **1992**, *13*, 65–72. [CrossRef]
55. Wensink, W. Dielectric properties of wet soils in the frequency range 1–3000 mhz. *Geophys. Prospect.* **1993**, *41*, 671–696. [CrossRef]
56. Wagner, N.; Emmerich, K.; Bonitz, F.; Kupfer, K. Experimental investigations on the frequency- and temperature-dependent dielectric material properties of soil. *IEEE Trans. Geosci. Remote Sens.* **2011**, *49*, 2518–2530. [CrossRef]
57. Lasne, Y.; Paillou, P.; Freeman, A.; Farr, T.; McDonald, K.C.; Ruffie, G.; Malezieux, J.M.; Chapman, B.; Demontoux, F. Effect of salinity on the dielectric properties of geological materials: Implication for soil moisture detection by means of radar remote sensing. *IEEE Trans. Geosci. Remote Sens.* **2008**, *46*, 1674–1688. [CrossRef]
58. Robinson, D.; Gardner, C.; Evans, J.; Cooper, J.; Hodnett, M.; Bell, J. The dielectric calibration of capacitance probes for soil hydrology using an oscillation frequency response model. *Hydrol. Earth Syst. Sci.* **1998**, *2*, 111–120. [CrossRef]

59. Dean, T.; Bell, J.; Baty, A. Soil moisture measurement by an improved capacitance technique, part I. Sensor design and performance. *J. Hydrol.* **1987**, *93*, 67–78. [CrossRef]
60. Fares, A.; Polyakov, V. Advances in crop water management using capacitive water sensors. In *Advances in Agronomy*; Academic Press: New York, NY, USA, 2006; Volume 90, pp. 43–77.
61. Standards Australia. Methods of testing soils for engineering purposes. In *Method 3.6.1: Soil Classification Tests—Determination of the Particle Size Distribution of a Soil—Standard Method of Analysis by Sieving*; Standards Australia Limited: Sydney, Australia, 2009; p. 9.
62. Standards Australia. Methods of testing soils for engineering purposes. In *Method 3.6.3: Soil Classification Tests—Determination of the Particle Size Distribution of a Soil—Standard Method of Fine Analysis Using a Hydrometer*; Standards Australia Limited: Sydney, Australia, 2003; p. 18.
63. Standards Australia. Methods of testing soils for engineering purposes. In *Method 3.2.1: Soil Classification Tests—Determination of the Plastic Limit of a Soil—Standard Method*; Standards Australia Limited: Sydney, Australia, 2009.
64. Standards Australia. Methods of testing soils for engineering purposes. In *Method 3.9.1: Soil Classification Tests— Determination of the Cone Liquid Limit of a Soil*; Standards Australia Limited: Sydney, Australia, 2015.
65. American Society for Testing and Materials. *Standard Test Methods for Organic Matter Content of Athletic Field Rootzone Mixes*; ASTM International: West Conshohocken, PA, USA, 2011.
66. Narsilio, G.A.; Disfani, M.M.; Orangi, A. Discussion of "fines classification based on sensitivity to pore-fluid chemistry" by junbong jang and j. Carlos santamarina. *J. Geotech. Geoenviron. Eng.* **2017**, *143*, 07017009. [CrossRef]
67. Santamarina, J.C.; Klein, K.; Fam, M. *Soils and Waves: Particulate Materials Behavior, Characterization and Process Monitoring*; Wiley: New York, NY, USA, 2001.
68. Standards Australia. *Geotechnical Site Investigations*; Standards Australia Limited: Sydney, Australia, 2009.
69. Agriculture Victoria. *Salinity Indicator Plants—A Guide to Spotting Soil Salting*; Agriculture Victoria: Victoria, Australia, 2018.
70. Wagner, N.; Schwing, M.; Scheuermann, A. Numerical 3-d FEM and experimental analysis of the open-ended coaxial line technique for microwave dielectric spectroscopy on soil. *IEEE Trans. Geosci. Remote Sens.* **2014**, *52*, 880–893. [CrossRef]
71. Orangi, A.; Narsilio, G.A. New capacitive sensor for in-situ soil moisture estimation. In Proceedings of the XV Pan-American Conference on Soil Mechanics and Geotechnical Engineering, Buenos Aires, Argentina, 15–18 November 2016; Manzanal, D., Sfrsio, A.O., Eds.; IOS Press: Buenos Aires, Argentina, 2016; pp. 422–429.
72. Dong, X.; Wang, Y.-H. The effects of the pH-influenced structure on the dielectric properties of kaolinite–water mixtures. *Soil Sci. Soc. Am. J.* **2008**, *72*, 1532. [CrossRef]
73. Henok, H.; Dinesh, S.; Frank, W.; Norman, W. Thermal and dielectric behaviour of fine-grained soils. *Environ. Geotech.* **2016**, *4*, 79–93.
74. Keysight Technologies (Ed.) *Basics of Measuring the Dielectric Properties of Materials*; Keysight Technologies: Santa Rosa, CA, USA, 2015.
75. Analog Devices (Ed.) *24-Bit Capacitance-to-Digital Converter with Temperature Sensor (ad7745/ad7746)*; Analog Devices: Boston, MA, USA, 2005.
76. Freescale Semiconductor Inc. *Mpr 121 Proximity Capacitive Touch Sensor Controller*; Freescale Semiconductor Inc.: Austin, TX, USA, 2010.
77. Microchip Technology Incoeporated. *Pic12(l)f1840 Data Sheet*; Microchip Technology Incoeporated: Springs, CO, USA, 2011.
78. Standards Australia. Methods of testing soils for engineering purposes. In *Method 5.1.1: Soil Compaction and Density Tests—Determination of the Dry Density/Moisture Content Relation of a Soil Using Standard Compactive Effort*; Standards Australia International Ltd.: Sydney, Australia, 2003.
79. Standards Australia. Methods of testing soils for engineering purposes. In *Method 2.1.1: Soil Moisture Content tests—Determination of the Moisture Content of a Soil—Oven Drying Method (Standard Method)*; Standards Australia Limited: Sydney, Australia, 2005; p. 5.
80. Orangi, A.; Withers, N.M.; Langley, D.S.; Narsilio, G.A. In-situ soil water content and density estimations using new capacitive based sensors. In Proceedings of the 5th International Conference on Geotechnical and Geophysical Site Characterisation, Gold Coast, QLD, Australia, 5–9 September 2016.

81. Ulaby, F.T.; Moore, R.K.; Fung, A.K. *Microwave Remote Sensing: Active and Passive*; Advanced Book Program/World Science Division, 1981–1986; Addison-Wesley Pub. Co.: Reading, MA, USA, 1981.

82. Mironov, V.L.; Kosolapova, L.G.; Fomin, S.V. Physically and mineralogically based spectroscopic dielectric model for moist soils. *IEEE Trans. Geosci. Remote Sens.* **2009**, *47*, 2059–2070. [CrossRef]

83. Kelleners, T.; Paige, G.; Gray, S. Measurement of the dielectric properties of wyoming soils using electromagnetic sensors. *Soil Sci. Soc. Am. J.* **2009**, *73*, 1626–1637. [CrossRef]

84. Hallikainen, M.T.; Ulaby, F.T.; Dobson, M.C.; El-rayes, M.A.; Wu, L.k. Microwave dielectric behavior of wet soil-part 1: Empirical models and experimental observations. *IEEE Trans. Geosci. Remote Sens.* **1985**, *GE-23*, 25–34. [CrossRef]

85. Dobson, M.C.; Ulaby, F.T.; Hallikainen, M.T.; El-rayes, M.A. Microwave dielectric behavior of wet soil-part II: Dielectric mixing models. *IEEE Trans. Geosci. Remote Sens.* **1985**, *GE-23*, 35–46. [CrossRef]

86. Lauer, K.; Albrecht, C.; Salat, C.; Felix-Henningsen, P. Complex effective relative permittivity of soil samples from the taunus region (Germany). *J. Earth Sci.* **2010**, *21*, 961–967. [CrossRef]

87. Salat, C.; Junge, A. Dielectric permittivity of fine-grained fractions of soil samples from eastern spain at 200 mhz. *Geophysics* **2010**, *75*, J1–J9. [CrossRef]

88. Roth, C.; Malicki, M.; Plagge, R. Empirical evaluation of the relationship between soil dielectric constant and volumetric water content as the basis for calibrating soil moisture measurements by TDR. *J. Soil Sci.* **1992**, *43*, 1–13. [CrossRef]

89. Roth, K.; Schulin, R.; Flühler, H.; Attinger, W. Calibration of time domain reflectometry for water content measurement using a composite dielectric approach. *Water Resour. Res.* **1990**, *26*, 2267–2273. [CrossRef]

90. Orangi, A.; Narsilio, G. Physical characterisation of soils recovered from the anzac battlefield. *Near Surf. Geophys.* **2017**, *15*, 85–101.

91. Terry, A. Brase. *Precision Agriculture*; Cengage Learning, Inc.: New York, NY, USA, 2005.

![sensors](sensors logo)

sensors

MDPI

Article

Consumer Grade Weather Stations for Wooden Structure Fire Risk Assessment

Torgrim Log

Department of Fire Safety and HSE Eng., Glöð R&D, Western Norway University of Applied Sciences, 5528 Haugesund, Norway; torgrim.log@hvl.no; Tel.: +47-900-500-01

Received: 14 August 2018; Accepted: 25 September 2018; Published: 27 September 2018

Abstract: During January 2014, Norway experienced unusually cold and dry weather conditions leading to very low indoor relative humidity (RH) in inhabited (heated) wooden homes. The resulting dry wood played an important role in the two most severe accidental fires in Norway recorded since 1923. The present work describes testing of low cost consumer grade weather stations for recording temperature and relative humidity as a proxy for dry wood structural fire risk assessment. Calibration of the weather stations relative humidity (RH) sensors was done in an atmosphere stabilized by water saturated LiCl, MgCl₂ and NaCl solutions, i.e., in the range 11% RH to 75% RH. When calibrated, the weather station results were well within ±3% RH. During the winter 2015/2016 weather stations were placed in the living room in eight wooden buildings. A period of significantly increased fire risk was identified in January 2016. The results from the outdoor sensors compared favorably with the readings from a local meteorological station, and showed some interesting details, such as higher ambient relative humidity for a home close to a large and comparably warmer sea surface. It was also revealed that a forecast predicting low humidity content gave results close to the observed outdoor weather station data, at least for the first 48 h forecast.

Keywords: relative humidity; consumer grade weather stations; calibration; winter fire risk

---

1. Introduction

Fire is a major cause of accidental injury and results in over 300,000 deaths annually [1–3]. The fires are usually associated with combustible spill accidents and fires in hot climates. Recently, subzero-temperature fires have, however, caught increased attention from researchers who have found these fires to be extremely severe and fast developing [4]. The Lærdalsøyri fire in Western Norway 18–19 January 2014 destroying 40 buildings and threatening the whole village including the historical Old Lærdalsøyri [5], may serve as an example. Ten days later, the Flatanger wild fire resulted in the loss of 60 structures in the Trøndelag region, a short distance south of the Arctic Circle [6].

In these areas, as well as in the majority of the country, wooden homes dominate the building style due to the abundance of wood as a construction material. One of the precursors for the severe fires was ambient subzero temperature air of low relative humidity for a few weeks before these fires. This resulted in dry outdoor combustible materials and extremely dry indoor wood in inhabited (heated) wooden structures [5,6]. A correlation between urban building fire frequency and low dew point temperature during winter time for selected areas in the USA was demonstrated as early as in 1956 by Pirsko and Fons [7]. It is also generally known that building fires are more common during winter in cold climates [8]. To monitor the temperature and the ambient relative humidity outdoor, as well as indoor, may then represent an indirect way of monitoring gradually changing winter fire risk. The wind plays a major role in the fire spread between buildings. The weather forecasts for the next days in combination with the monitored indoor RH may then in the future probably be used to design a conflagration fire danger rating system [9].

However, such an application reveals the need for low cost sensors for recording ambient and indoor temperature and RH. Recently, a fast TiO_2 capacitive high speed sensor with 35 s response time [10] and an extremely fast response induced stress-optic polymer fiber sensors with only 2 s response time [11] have been developed. The relative humidity indoors does, however, change slowly. A one to two hour response time would be satisfactory at least for testing the concept in the present study. To assist in data collection, it would be preferable to use equipment that allows for remote access of the recorded data through the internet for analysis and risk evaluation. Possible warnings to the fire brigades in case of unusual dry conditions developing could then be issued. It was therefore considered to determine whether comparably low cost consumer grade weather stations could be used for relative humidity measurements as a proxy for increased fire risk.

The objective of the present work is to report on the experience of testing consumer grade weather stations for relative humidity measurements as a proxy for the winter fire risk in wooden constructions (Section 1). Section 2 presents the theoretical background. Section 3 describes the need for calibration and a simple way to establish calibration curves without a professional climate chamber. Section 4 presents the results regarding the main objective as well as some other interesting findings. In Section 5 the overall experience with the consumer grade weather stations for scientific measurements are discussed. Suggestions for improvement and future research are also presented.

2. Theory

2.1. The Relative Humidity Conten in Air

The relative humidity in air is dependent on the absolute water vapor content as well as the air temperature. The saturation vapor pressure of water is a near exponential function of temperature and may be described by [12]:

$$P_{sat} = 610.78 \cdot \exp\left(\frac{17.2694 \cdot T_o}{T_o + 238.3}\right) (Pa), \tag{1}$$

where T_o (°C) is the ambient temperature. When the temperature and the relative humidity of the air are known, the water vapor concentration may be calculated by:

$$C_{w,o} = RH_o \cdot \frac{P_{sat} \cdot M_w}{R \cdot T_o}, \left(kg\ m^{-3}\right), \tag{2}$$

where RH_o is the ambient relative humidity (in the range 0 to 1.0), M_w (0.01802 kg mol^{-1}) is the molecular mass of water and R (8.314 J K^{-1} mol^{-1}) is the molar gas constant.

Cooling this air so that it theoretically contains more than 100% relative humidity, the surplus water condenses at small dust particles, etc. to make fog. It is a quite normal phenomenon to observe fog in a cold night and morning dew on the grass after a cold night. It should be noted that cooling the air results in a volume reduction, which also increases the water vapor concentration. The focus of the present paper is, however, the indoor heating of the air in cold climates. Heating the cold ambient air (at constant pressure) to the higher indoor temperature, T_{in} (°C), will, according to the ideal gas law, result in a dilution corresponding to the gas volume expansion, i.e.,

$$C_{w,in} = C_{w,o} \cdot \left(\frac{T_o + 273.15\ K}{T_{in} + 273.15\ K}\right) \left(kg\ m^{-3}\right). \tag{3}$$

2.2. Concentration of Water in the Air

In Norway, historic weather data at 1 h frequency may be retrieved for free from a number of meteorological stations by the application provided at www.eklima.no. In the Haugesund area, the meteorological station at the Haugesund airport is well equipped and professionally maintained for recording data relevant for the present study. Based on retrieved temperature and relative humidity data, the historic values for the ambient water vapor concentration may be calculated by using Equations (1) and (2).

The calculated outdoor water concentration based on data from the meteorological station at the Haugesund airport for the period July 2015 through June 2016 is shown in Figure 1. The variation through the year is significant. With a few exceptions, it is seen that the low water content in parts of January, i.e., about 2 g/m^3, was indeed very low compared to the rest of the year. If such dips are short, i.e., less than a day, a wooden home does not have time to adjust to the new dry conditions. This is partly due to the finite air exchange rate as well as slow diffusion processes of water in wood and humidity being released from the drying wood to the indoor air. However, if the dry conditions persist for some days or weeks, the wood dries out and the fire risk related to the drier wood increases [5]. In order to analyze this situation, information about the indoor relative humidity is needed. Preferably, this should be recorded in a cost efficient and convenient way, e.g., through the internet.

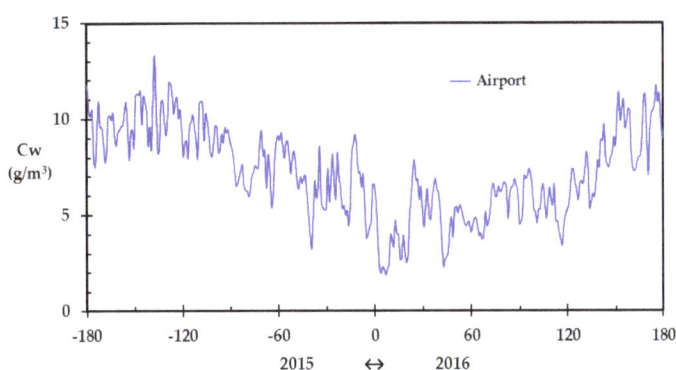

Figure 1. Calculated outdoor water concentration based on temperature and relative humidity recorded by the meteorological station (Haugesund airport), July 2015 through June 2016. (The x-axis represents days prior to and after New Year).

2.3. Moisture Supply Sources Indoors

Indoors, there are usually also sources of moisture supply present, e.g., humans, pets and pot plants. Hygroscopic materials, e.g., wood and other cellulose based materials such as upholstery, clothes, carpets, etc. may also release humidity when the indoor climate gets drier. If the indoor air gets more humid, these materials absorb humidity from the air. These materials may also show hysteresis when changing from the mode of adsorption to desorption and vice versa [13]. Since such hysteresis is dependent on parameters such as thickness, internal humidity diffusion processes, etc. of the variety of materials involved, it is quite complicated to model the contribution of these materials to the indoor humidity levels.

The best way to measure the water content of indoor combustible hygroscopic materials would probably be to record the mass change of selected objects. This is not easy to do with a good precision. Measuring the indoor relative humidity does, however, provide a good alternative for obtaining information about the indoor climate contribution to structural wood fire risk. Recording the humidity levels, and keeping track of especially low indoor humidity levels, gives a good indication about the structural wood fire risk development. Warnings may be considered based on weather forecasts so that focus can be shifted to monitoring indoor relative humidity development and fire risk evaluations when needed.

3. Materials and Methods

3.1. Initial Considerations and Preliminary Testing

The first idea was to record the mass of representative indoor wooden objects in order to consider their combustibility. This could be e.g., wooden plates of different thicknesses placed on separate balances, and then record the mass at a convenient frequency. The recorded data could then be transferred via the internet to researchers analyzing the data. It was, however, realized that it would be quite challenging to e.g., have sufficiently stable balances. A number of precision balances would also have been very costly. During discussions with an automation engineer [14], it was decided to try consumer grade weather stations for recording relative humidity as a proxy for the indoor wood fire risk. It was therefore decided to abandon the direct mass recordings and further pursue the weather station proposal.

A very low cost, and not web based, weather station was therefore purchased and evaluated against a precision psychrometer (Extec RH390, ID 10116891, calibration certificate 350342-10116891, Extech Instruments, Waltham, MA, USA). The psychrometer calibration data are presented in Table 1. A straight line was fit to the RH390 psychrometer RH readings as a function of the climate chamber RH to give:

$$RH = 1.0345 \cdot RH_{reading} - 0.5683, \tag{4}$$

where the deviations are stated in the last column of Table 1.

Table 1. Producer calibration data for the Extec RH390 precision psychrometer (in a Binder KBF-115 climate chamber, serial No. 09-06299, Binder GmbH, Ulm, Germany, traceable to Scalibra AS, Skjetten, Norway, proof No. 3491-14).

RH_{real}	$RH_{reading}$	Equation (4) Deviation
24.0%	23.7%	−0.05%
49.5%	48.5%	0.10%
75.0%	73.0%	−0.05%

The weather station and the precision psychrometer were placed in a 17 L transparent plastic box with a lid. Operation of the precision psychrometer without disturbing the internal atmosphere was provided for by a 2.5 mm plastic rod through a sealed hole. The internal relative humidity was adjusted to get varying relative humidity (RH) in the plastic box. To get a low RH, the box was taken outdoor and filled with low temperature air which became dry when adjusted to indoor conditions at 22 °C. To get high RH, droplets of water were allowed to evaporate in the plastic box, inspired by a previous study [15]. Comparing the results recorded by the very inexpensive weather station with the results obtained by the precision psychrometer, a simple straight line, similar to the one described by Equation (4), was obtained for the deviation. Subsequent recordings corrected by this straight line gave results within the psychrometer precision, i.e., within 2% RH. The response time and longtime stability of this inexpensive weather station were not further investigated. The results obtained were, however, very promising regarding the potential of using more advanced consumer grade weather stations. It was therefore decided to proceed with a higher quality weather station, which had a built in system for web based data transfer.

3.2. The Netatmo Weather Stations

Eight Netatmo weather stations (Netatmo Urban Weather Station, Wi-Fi, NWS01-EC, [16]), as shown in Figure 2, were purchased from a local hardware shop. These units were set up to the local WiFi network by a PC for later data access by smartphones and PCs through a password protected WiFi connection. These weather stations record temperature and relative humidity at 5 min intervals. The indoor unit also records atmospheric pressure and CO_2-concentration. The outdoor unit sends the recorded data to the indoor unit, which automatically forwards the recorded indoor and outdoor data to the Netatmo servers for instant or later user access and data retrieval. The complete weather station, including indoor and outdoor units, is shown in Figure 1. The indoor units are powered by a USB adapter while the outdoor units are battery powered. The outdoor sensor battery status is transmitted to the user web page. It is also possible to give public access to temperature and RH recordings. This service is currently being used to improve the temperature predictions of the Norwegian weather forecasts [17].

The producer of the Netatmo units states that the equipment can read temperatures within ±0.5 °C and relative humidity (RH) within ±3% RH. The PC interface allows for a temperature calibration by the user. In the present work it was decided to calibrate the temperature output of the units against a precision temperature instrument (ALMEMO® SP10302D Pt100 Temperature Reference Instrument, Ahlborn Mess- und Regelungstechnik GmbH, Holzkirchen, Germany) for the temperature interval relevant for the indoor and outdoor sensors. Only minor adjustments were needed to achieve temperature recordings well within ±0.3 °C for the temperature region of interest in the present work.

Figure 2. Netatmo weather station outdoor unit (**left**) and indoor unit (**right**). (Photo by Netatmo, reproduced with permission).

There was, however, no web interface available for the user to calibrate the RH readings. The producer of the Netatmo units only provides a single point adjustment at about 75% RH. When testing the readings of two Netatmo indoor and outdoor units against the precision psychrometer, a clear need for calibrating the weather station indoor and outdoor readings was revealed. It was therefore decided to obtain calibration curves for all indoor and outdoor units.

3.3. The Equipment and Chemicals Used for RH Calibration

To make an atmosphere of known and constant temperature and relative humidity (RH), the sensors to be calibrated were placed, four at a time, in a transparent plastic box of dimensions 50 cm by 40 cm by 30 cm height. A lid was applied to the box and the USB power cables for the indoor units were let out through the back wall. The hole was sealed to prevent air exchange with the surroundings.

Water saturated inorganic salt solutions, i.e., LiCl, $MgCl_2$ and NaCl, all salts pro analysis quality, were used to respectively achieve RH values of 11.3 ± 0.31% RH, 33.07 ± 0.18% RH and 75.47 ± 0.14% RH [18]. This range covered the area of most interest for the indoor sensors, i.e., 15–50% RH and the outdoor sensors, i.e., above 40% RH. Distilled water was applied to the inorganic salts to minimize the influence of any impurities. The sensor readings during the calibration were obtained at 2–5 °C and 22 °C, for the outdoor and indoor units, respectively.

Saturated water salt solutions were placed at four 8 cm diameter plastic plates. Two of these plates were placed on the plastic box floor level and two plates were placed at about 10 cm elevation. This was done in order to quickly obtain equilibrium RH conditions after any handling of salts or weather station units in the plastic box. Air circulation within the box was achieved using a USB fan, which also speeded up the process of establishing a constant RH level. When changing to NaCl solutions, water droplets were added to an aluminum plate (for good heat transfer to the droplets [19]) placed just downstream the USB fan to quickly achieve a humid atmosphere.

The previously mentioned battery powered precision psychrometer was also placed inside the plastic container for instant recordings of temperature and relative humidity. As with the preliminary testing, a plastic rod was arranged for operating the psychrometer. The psychrometer readings were observed through the transparent plastic box lid. The box was loosely covered by sheets of aluminum foil to reflect light radiation to prevent any plastic box "greenhouse effects". The system was then left to achieve equilibrium conditions governed by the applied saturated salt solution. The psychrometer readings were used to confirm that a constant RH level was achieved inside the plastic box. Another two hours were allowed for ensuring constant weather station readings. The temperature and RH for each weather station unit were then obtained by the web interface together with the psychrometer readings. The procedure was repeated for all the relative humidity sensors (four at a time) and for all the saturated inorganic salt solutions. Separate testing was done to establish the sensor response times when exposed to a sudden change in relative humidity.

4. Results

4.1. Calibration Coefficients for the Weather Station Relative Humidity Recordings

The sensors generally gave very erroneous results compared to the equilibrium conditions ensured by the water saturated salt solutions. Several of the detectors were up to, and even more than, 10% RH wrong either at low (11.3% RH) or high relative humidity (75.5% RH). The sensor outputs were, however, very linear with respect to the real relative humidity. This indicated that a linear equation could be used for converting the recorded result to correct relative humidity values. A linear correction curve, similar to the one shown in Equation (4), was therefore fit to the RH data recorded for the indoor units (at 22 °C) and outdoor units (at 2–5 °C). The results are presented in Table 2. The regression coefficients were better than 0.995 for all the units indicating a good straight line fit. However, when studying the slopes and intercepts presented in Table 2, it is quite clear that calibration was indeed necessary for all the sensors.

Table 2. The linear correction curve for the eight weather stations studied.

Sensor	Slope	Intercept
Indoor 1	1.030	−7.778
Outdoor 1	1.092	−6.334
Indoor 2	1.074	−10.811
Outdoor 2	1.090	−6.734
Indoor 3	1.122	−16.707
Outdoor 3	1.108	−7.464
Indoor 4	1.045	−11.716
Outdoor 4	1.118	−8.694
Indoor 5	0.991	−7.713
utdoor 5	1.107	−7.973
Indoor 6	0.935	−4.655
Outdoor 6	1.1051	−4.826
Indoor 7	0.999	−7.872
Outdoor 7	1.103	−6.420
Indoor 8	0.989	−8.778
Outdoor 8	1.138	−9.091

It should be noted that when extrapolating the linear curves based on the correction coefficients presented in Table 2, for some of the sensors a recorded 100% RH would turn out to yield corrected values slightly above 100% RH. This was solved by forcing any corrected result above 100% RH to 100% RH.

A number of tests were also done in the range 40% to 60% RH, where the precision psychrometer, corrected by its calibration curve, was used as the reference. No deviation from linearity was detected for any of the weather station units. To the surprise of the author, when testing an outdoor unit at indoor temperatures, the results were still within 2% RH. The calibration of the outdoor sensors at 2 °C to 5 °C was therefore taken as valid also for temperatures below 0 °C.

4.2. Relative Humidity Sensor Response Times

The sensor response time was also tested. This was done by setting the weather station number 1 indoor and outdoor units in an atmosphere stabilized by water saturated NaCl solution for 6 h. Then, the sensors were suddenly taken out and placed in the measurement container where the atmosphere was in equilibrium with water saturated LiCl solution. It should, however be mentioned that this does not represent a clean cut step function from humid (75% RH) atmosphere to dry atmosphere (11% RH) since during the sensor handling, the air in the plastic box chamber was partly diluted by indoor air of about 35% RH. Nevertheless, compared by the recording time of several hours, given the experimental setup this was as close to a step function as practicably possible. The results are shown in Figure 3, normalized such that a result of 1.0 corresponds to reaching the new constant value. The sensors did not show any change after about 3.5 h. The outdoor unit reached 95% change within 1 h while the indoor results reached 95% change towards the new constant value at just above 2 h, Whether this 2 h 95% response time was sufficient for the present work was pending the results for the measurements in real buildings, where the changes in indoor relative humidity over time was assumed to be much slower than 2 h.

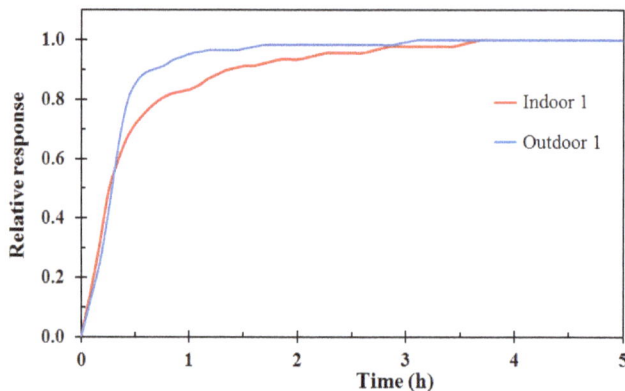

Figure 3. Weather station number 1 indoor (red) and outdoor (blue) sensors relative response to a near step change in relative humidity from 75% RH to 11% RH.

4.3. Results Regarding Winter Fire Risk

The most interesting period during the winter 2015/2016 was in January 2016. Early in January, the wind direction changed and the wind came from the mountain plains east of Haugesund, Norway. This led to adiabatically heated ambient air and lower than normal outdoor relative humidity. The recorded temperature and relative humidity from the local meteorological station at Haugesund airport for January 2016 is presented in Figure 4.

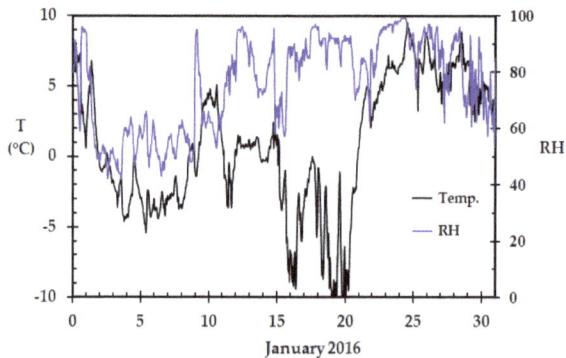

Figure 4. Recorded temperature and relative humidity at the local meteorological station (Haugesund airport), January 2016.

The adiabatically heated dry subzero temperature air is easily identified in the first days of January. However, as both the temperature and the relative humidity vary, it is easier to interpret the data when studying the actual ambient air water concentration.

The water vapor concentration calculated by Equations (1) and (2), based on the Netatmo weather stations outdoor temperature and relative humidity at two homes, as well as the Haugesund airport, is shown in Figure 5. It is clearly seen from Figure 5 that the recordings at the two homes agreed very well with the recordings at the local meteorological station. This was also the case for the other six, but one, of the Netatmo weather stations. The weather station results that deviated, and a possible explanation for this deviation, are presented in Section 4.5.

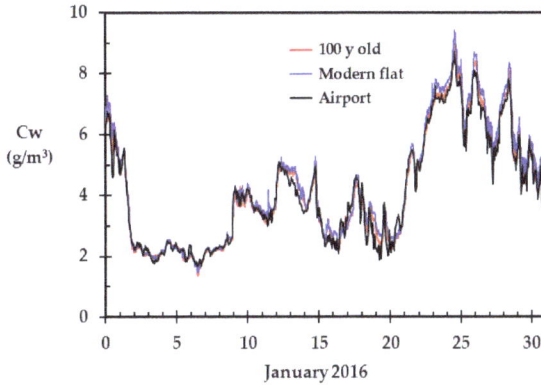

Figure 5. Calculated outdoor water vapor concentration for the recordings outside the 100 years old home, the modern flat and the local meteorological station (Haugesund airport), January 2016.

The indoor relative humidity recorded at the homes presented in Figure 5, is shown in Figure 6 during the January 2016 cold snap. It is seen that the indoor relative humidity was as low as 20% RH in periods before it started to increase significantly from 21 January. It is also seen that there are some indoor relative humidity peaks in both homes. Since these peaks are not synchronized in time, they are probably a result of indoor human activity.

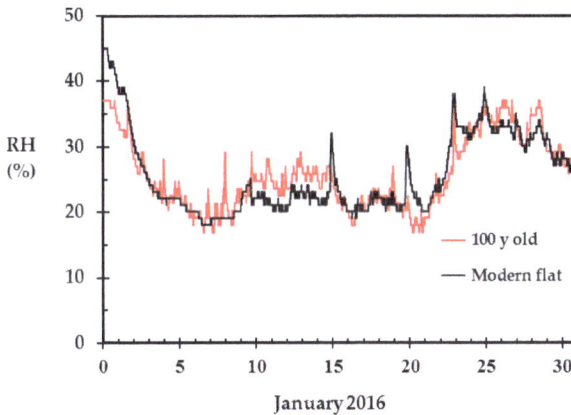

Figure 6. Recorded indoor RH for a 100 years old wooden home (inhabited by four persons) and a modern flat (inhabited by two persons) in Haugesund, Norway, during the cold snap in January 2016.

Especially during winter time, with large outdoor to indoor temperature differences, there will be, and should be, sufficient air changes in a building to prevent rot formation in colder parts of the thermal insulation, etc. Due to the chimney effect, the indoor air will gradually be exchanged with ambient air. The ambient air entrained into the home is then heated to the indoor temperature. Wind pressure also increases the air change rate. Indoors, there will usually be some moisture gain in inhabited buildings as a result of persons and pets breathing or perspiring, dishwashing, pot plants and wood sorption processes. The wood sorption processes may go both ways, i.e., represent a positive or negative humidity gain. This also holds for other hygroscopic materials, such as furniture upholstery, pot plant soil, etc. However, assuming that there is no moisture gain, and that the ambient air is heated to the indoor conditions, a theoretical indoor relative humidity may be calculated based on the ambient conditions. The indoor weather station unit supplies the indoor temperature needed for the calculations while the ambient temperature and RH may be taken from the outdoor unit or from the local meteorological station. The best way is, however, to record the indoor RH directly, as seen in Figure 6.

It has recently been demonstrated in $\frac{1}{4}$ scale wooden compartments that the time for reaching flashover, i.e., the sudden transition from a gradually increasing fire to a fully engulfing compartment fire, was very dependent on the wood fuel moisture content (FMC). At equilibrium conditions at 20% RH, the wood equilibrium moisture content (EMC) is about 4.5% [5]. At 50% RH the EMC is about 9.3%. The $\frac{1}{4}$ scale fire testing revealed that the time to flashover when the FMC was 4.5% was just short of half the time needed to reach flashover at 9.3% FMC [19]. This dramatically reduces the margins for escape and rescue, and may in some situations be extremely critical and lead to major loss of lives [20]. Recordings like the one presented in Figure 6 may help identifying these critical situations so that proper mitigating measures may be taken for risk management.

4.4. Weather Forecast Data

31 December 2015, at 23:00, a weather forecast API from the Norwegian Meteorological Institute was run to retrieve the forecasts for the following 10 days period. For the first 48 h, the data are detailed to each hour. Then, the data is gradually presented with less frequency. Five days later, the API was run again supplying data for the following 10 days period. The calculated ambient air water concentration based on the forecasted data is presented in Figure 7 together with the water concentration based on the subsequent recordings from the local meteorological station, i.e., the Haugesund Airport. It is evident from Figure 7 that the forecasts are quite good, at least for the first 48 h. This is very interesting when it comes to potential future predictions of the dry wood fire risk based merely on weather forecasts, as recently suggested [9]. When becoming alarmed by the weather forecast, the focus can be shifted to following up closely on selected homes with calibrated in-house weather stations, as e.g., presented in Figure 6, to confirm an increasing wooden home fire risk.

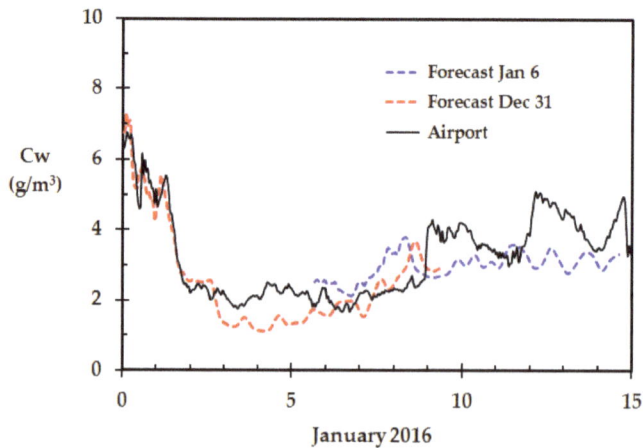

Figure 7. Calculated outdoor water concentration based on the forecasts at 31 December and 6 January and data subsequently retrieved from the meteorological station at Haugesund airport.

4.5. Other Results of Interest

The only outdoor weather station sensor results that differed significantly from those obtained by the other sensors were from the home in Skudeneshavn, see Figure 8. This was especially the case during the first 12 days of January 2016. From 1 January to 12 January, the wind direction was about 80–100°, i.e., easterly wind from the central south Norwegian mountain plains. 12 January, the wind direction changed to about 180°, i.e., southerly wind from the open sea.

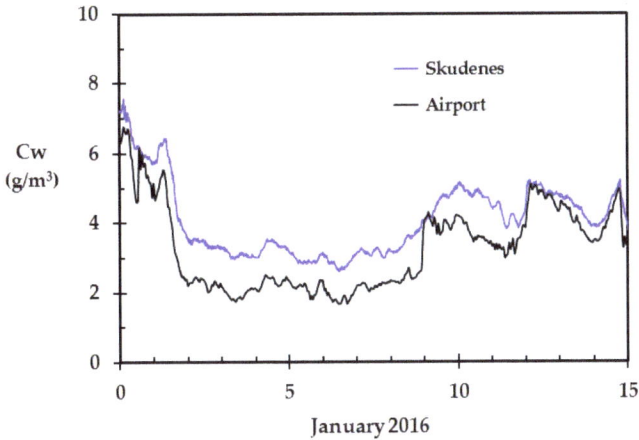

Figure 8. Calculated outdoor water concentration based on data from the meteorological station at Haugesund airport and data from the weather station in Skudeneshavn.

While the other homes studied were close to the town of Haugesund, the one standing out was in Skudeneshavn, i.e., on the southern tip of the Karmøy island. As can be seen in middle left part of Figure 9, this location is more or less surrounded by sea water.

Figure 9. Map of the fjord Boknafjorden between Haugesund and Stavanger, Norway. North up on the map. (Accessible for free from www.norgeibilder.no).

The major fjord Boknafjorden, between Haugesund and Stavanger, has open water the whole year while the inner parts of the attached smaller fjords are covered by ice in January. In wind from east, the air passing over the open sea has a long fetch and therefore a long contact time for picking up humidity from the open water surface before arriving at Skudeneshavn. This is especially the case during winter time when the sea has a temperature well above the ambient air temperature, and when the ambient air from the east is dry due to adiabatically heating. The reason for the unit in Skudeneshavn showing the highest RH values is therefore most likely due to the dry wind from the east picking up humidity on this 25+ km long fetch during the first 12 days of January 2016.

The smaller fjord arms and most upwind lakes east of Haugesund were covered by ice during January 2016. The other weather stations close to the town of Haugesund, as seen in the upper left part of Figure 9, were therefore in the period 1 January to 12 January exposed to air that had not travelled over open and comparably warm water surfaces. When the wind direction changed on 12 January, and thereafter came from the south, all the weather stations including the one at Haugesund airport were exposed to the humid air from the sea.

5. Discussion

The purpose of this study was to test whether consumer grade weather stations for recording temperature and relative humidity could be used for indicating structural wooden home fire risk during winter time. The present work demonstrated that calibration was needed for the relative humidity sensor results to be within ±3% RH. The correction curves were simple linear equations. Without calibration, the readings were quite erroneous, i.e., the errors in relative humidity were up to more than 10% RH.

The calibration was done in a transparent plastic box where the air was conditioned by LiCl, $MgCl_2$ and NaCl pa salt saturated water solutions. Internal air circulation was ensured by an USB powered fan. This represented a low cost setup for the relative humidity calibration. An available high precision temperature unit was used to calibrate the weather station by the built-in user based temperature calibration app. The recorded temperatures were well within 0.3 °C when tested after this user calibration. Using the high precision temperature unit may, however, be seen as an overkill since a normal type K thermocouples would probably have been sufficiently accurate for the temperatures of interest in present work.

The precision psychrometer was used to check that the inorganic salts did indeed produce the correct relative humidity. This was also in principle not necessary, as the saturated salt solutions would ensure equilibrium conditions at a known relative humidity level [18]. The psychrometer did, however, represent an independent way for checking that the correct saturated salt solution was applied and gave a good indication about the time needed to achieve the actual equilibrium conditions.

The selected weather stations had a built-in web application which allowed for remote retrieval of the data recorded at 5 min intervals. The recorded data was downloaded in the form of spread sheets. It was therefore easy to apply the proper relative humidity correction equation for each indoor and outdoor detector unit.

The tested weather stations had a resolution of 1% RH. This made it difficult to detect any weather station hysteresis regarding the relative humidity recordings. The weather station response was slow, i.e., a time period of about 2 h was therefore allowed to ensure correct results. The sensors can therefore not be used in situations where a fast response time is required. When the ambient water vapor concentration decreased considerably in January 2016, it was seen that a typical home needed a few days to equilibrate to the new and drier condition, as seen in Figure 6.

In principle, the measurement frequency of 12 h^{-1} allows for detecting quite rapid changes in the indoor relative humidity, e.g., as a result of airing. However, given the slow response of the units, i.e., close to two hours, any such potential sudden changes in indoor relative humidity could not be discovered. Peaks due to cooking activities were, however, discovered. Due to the slow response, the peak values were probably not reliable as the sensor recordings were probably lagging behind such rather rapid changes in indoor relative humidity.

The weather station response time must be compared to the building response time, i.e., a few days. The conservative two hours weather station response time was an order of magnitude faster than the building indoor response time. This was taken as a proof of the sensor response time being sufficiently fast for the purpose of the present work, i.e., demonstrating that consumer grade weather stations can be used to monitor the dry indoor climate fire risk levels of wooden buildings.

It was clearly seen that the indoor sensor had a longer response time compared to the outdoor sensor when tested under the same conditions, i.e., 2 h versus 1 h to 95% of a step change in relative humidity. This may simply be due to the larger internal air volume of the indoor sensors. For future measurement campaigns it is recommended to try to optimize the response time by e.g., allowing for better air convection and thereby faster response.

During the calibration, the outdoor relative humidity sensors were, as always, powered by internal batteries. The indoor relative humidity sensors were, however, supplied USB power from transformers outside the box, i.e., the current wires had to be led through the wall. This was also the case for the USB fan power cable. Instead of relying on external power, using USB power banks for supplying these units would take away the clutter of wires through the wall. Utilizing USB power banks as the power supply source are therefore recommended for future calibration and characterization studies.

The accuracy of the recordings, after the calibration curves were established and applied to the retrieved data, was found to be within ±3% RH. This was also the case for three weather stations that were tested after four months of operation during December 2015 through March 2016. This accuracy was still well within the goal for the present study. Advanced relative humidity sensors may display results far better than this, and with response times of 35 s [10] or even as low as 2 s [11]. Such improved relative humidity sensors may become available on the consumer market. But for now, as no dry wood fire risk warning system exists, the weather stations tested in the present work was found fit for purpose. They did work quite well as an indicator of dry indoor climate resulting in a gradually increasing wooden home structural fire risk. The long time stability through winter and summer season was not tested in the present work since, based on the simplicity of the inorganic salt calibration, the detectors will regardless be recalibrated for future measurement campaigns.

An interesting research possibility regarding web based weather stations may be related to the geography teaching for primary and secondary schools. The influence of the sea and the mountains on the local relative humidity, as clearly observed in the present work, may be demonstrated in the class room. The changes in relative humidity when the temperature is changing may also be worthwhile presenting to school children. The calibration procedure for teaching purposes may be done without using pro analysis quality salts. Commercial grade NaCl is available in any grocery store, though maybe producing equilibrium conditions 1% RH off due to a minor content of other salts. Consumer grade $MgCl_2$ may be available at hardware stores as an ice melting powder, at least in cold climate areas. $MgCl_2$, which in contrast to LiCl, is not poisonous, may therefore represent a safe second calibration point.

Another research possibility may be to use private weather stations for suppling data for increased wildfire risk awareness, especially during summer time [21], but also during winter time in cold climate areas [6]. Recent wildland urban interface (WUI) fires in Europe have also claimed many lives, such as the one in Portugal 16–24 June 2017, which claimed 66 lives [21]. The Athens area WUI fires during 20–22 July 2018 resulted in more than 90 fatalities [22]. If the public becomes more aware of the overhanging wildfire and wildland urban interface fire risk by reading properly calibrated ambient relative humidity data, they may be better prepared for preventing igniting start fires as well as for an early evacuation if a wildfire starts in their region. Increasing the risk awareness using e.g., phone app warnings based on private weather stations may be a worthwhile future research project.

The main conclusion of the present study is that consumer grade weather stations, at least the type tested in the present study, have been shown to quite accurately measure the indoor and outdoor relative humidity when properly calibrated. This indicates that such weather stations could become central in a future attempt to develop a cold climate structural fire risk warning system.

Funding: The research received no external funding.

Acknowledgments: The valuable discussions with colleagues in the oil and gas industry are highly appreciated. It is much appreciated that the home owners allowed for installation of weather stations during the winter 2015/2016. Valuable comments by the two anonymous reviewers are also appreciated.

Conflicts of Interest: The author declares no conflict of interest.

References

1. World Health Organization. Health Statistics and Information Systems: Estimates for 2000–2015. Available online: http://www.who.int/healthinfo/global_burden_disease/estimates/en/index1.html (accessed on 1 August 2017).
2. World Health Organization. Injuries and Violence: The Facts. Available online: http://www.who.int/violence_injury_prevention/key_facts/en/ (accessed on 14 March 2018).
3. World Health Organization. Burn Prevention, Success Stories, Lessons Learned. Available online: http://www.who.int/violence_injury_prevention/publications/other_injury/burn_success_stories/en/ (accessed on 14 March 2018).
4. Metallinou, M.M.; Log, T. Health impacts of climate change-induced subzero temperature fires. *Int. J. Environ. Res. Public Health* **2017**, *14*, 814. [CrossRef] [PubMed]
5. Log, T. Cold climate fire risk; A case study of the Lærdalsøyri Fire, January 2014. *Fire Technol.* **2016**, *52*, 1825–1843. [CrossRef]
6. Log, T.; Thuestad, G.; Velle, L.G.; Khattri, S.K.; Kleppe, G. Unmanaged heathland—A fire risk in subzero temperatures? *Fire Saf. J.* **2017**, *90*, 62–71. [CrossRef]
7. Pirsko, A.R.; Fons, W.L. *Frequency of Urban Building Fires as Related to Daily Weather Conditions*; Forest and Range Experiment Station: Berkeley, CA, USA, 1956.
8. Rohrer-Mirtschink, S.; Forster, N.; Giovanoli, P.; Guggenheim, M. Major burn injuries associated with Christmas celebrations: A 41-year experience from Switzerland. *Ann. Burns Fire Disasters* **2015**, *28*, 71–75. [PubMed]
9. Metallinou, M.M.; Log, T. Cold Climate Structural Fire Danger Rating System? *Challenges* **2018**, *9*, 12. [CrossRef]
10. Liu, M.-Q.; Wang, C.; Kim, N.-Y. High-Sensitivity and Low-Hysteresis Porous MIMType Capacitive Humidity Sensor Using Functional Polymer Mixed with TiO_2 Microparticles. *Sensors* **2017**, *17*, 284. [CrossRef] [PubMed]
11. Leal-Junior, A.; Frizera-Neto, A.; Marques, C.; Pontes, M.J. Measurement of Temperature and Relative Humidity with Polymer Optical Fiber Sensors Based on the Induced Stress-Optic Effect. *Sensors* **2018**, *18*, 916. [CrossRef] [PubMed]
12. Tetens, O. Uber einige meteorologische Begriffe. *Zeitschrift fur Geophysik* **1930**, *6*, 297–309.
13. Salin, J.G. Inclusion of the sorption hysteresis phenomenon in future drying models. Some basic considerations. *Maderas Cienc. Tecnol.* **2011**, *13*, 173–182. [CrossRef]
14. Langelandsvik, G.M. (Equinor, Kårstø, Norway) Personal communication, 2015.
15. Log, T. Water droplets evaporating on horizontal semi-infinite solids at room temperature. *Appl. Therm. Eng.* **2016**, *93*, 214–222. [CrossRef]
16. Personal Weather Station. Available online: https://www.netatmo.com/en-US/product/weather/weatherstation (accessed on 18 July 2018).
17. The Norwegian Meteorological Institute. Private Weather Data in the Yr-Forecast. Available online: http://www.yr.no/artikkel/private-vaerdata-inn-i-yr-varselet-1.13963299 (accessed on 17 July 2018).
18. Greenspan, L. Humidity Fixed Points of Binary Saturated Aqueous Solutions. *J. Res. Natl. Bur. Stand.* **1977**, *81*, 89–96. [CrossRef]

Sensors **2018**, *18*, 3244

19. Kraaijeveld, A.; Gunnarshaug, A.; Schei, B.; Log, T. Burning rate and time to flashover in wooden $\frac{1}{4}$ scale compartments as a function of fuel moisture content. In Proceedings of the 8th International Fire Science & Engineering Conference, Windsor, UK, 4–6 July 2016; pp. 553–558.
20. Delâge, C. *Rapport du Commissaire aux Incendies du Québec*; Coroners Fire Investigation Report; L'Isle-Verte Incendie: Québec, QC, Canada, 2015.
21. Benfield, B. Companion Volume to Weather, Climate & Catastrophe Insight, Additional Data to Accompany the 2017 Annual Report. 2018. Available online: http://thoughtleadership.aonbenfield.com/Documents/20180124-ab-if-annual-companion-volume.pdf (accessed on 8 August 2018).
22. Norwegian National Broadcasting Corporation. Available online: https://www.nrk.no/nyheter/skogbranner-i-hellas-1.14137775 (accessed on 6 August 2018).

sensors

MDPI

Article

Humidity Measurement in Carbon Dioxide with Capacitive Humidity Sensors at Low Temperature and Pressure

Andreas Lorek [1,*] and Jacek Majewski [2]

[1] German Aerospace Center (DLR), Rutherfordstraße 2, 12489 Berlin, Germany
[2] Department of Automation and Metrology, Faculty of Electrical Engineering and Computer Science, Lublin University of Technology, 38A Nadbystrzycka Str., 20-618 Lublin, Poland; j.majewski@pollub.pl
* Correspondence: andreas.lorek@dlr.de; Tel.: +49-306-705-5390

Received: 27 June 2018; Accepted: 6 August 2018; Published: 9 August 2018

Abstract: In experimental chambers for simulating the atmospheric near-surface conditions of Mars, or in situ measurements on Mars, the measurement of the humidity in carbon dioxide gas at low temperature and under low pressure is needed. For this purpose, polymer-based capacitive humidity sensors are used; however, these sensors are designed for measuring the humidity in the air on the Earth. The manufacturers provide only the generic calibration equation for standard environmental conditions in air, and temperature corrections of humidity signal. Because of the lack of freely available information regarding the behavior of the sensors in CO_2, the range of reliable results is limited. For these reasons, capacitive humidity sensors (Sensirion SHT75) were tested at the German Aerospace Center (DLR) in its Martian Simulation Facility (MSF). The sensors were investigated in cells with a continuously humidified carbon dioxide flow, for temperatures between -70 °C and 10 °C, and pressures between 10 hPa and 1000 hPa. For 28 temperature–pressure combinations, the sensor calibration equations were calculated together with temperature–dependent formulas for the coefficients of the equations. The characteristic curves obtained from the tests in CO_2 and in air were compared for selected temperature–pressure combinations. The results document a strong cross-sensitivity of the sensors to CO_2 and, compared with air, a strong pressure sensitivity as well. The reason could be an interaction of the molecules of CO_2 with the adsorption sites on the thin polymeric sensing layer. In these circumstances, an individual calibration for each pressure with respect to temperature is required. The performed experiments have shown that this kind of sensor can be a suitable, lightweight, and relatively inexpensive choice for applications in harsh environments such as on Mars.

Keywords: capacitive humidity sensors; SHT75; carbon dioxide; humidity; Mars in-situ measurements; experimental simulation chambers; Martian atmosphere; low temperature; low pressure; CO_2

1. Introduction

The exploration of Mars has become of growing importance in view of its relatively promising Martian environmental conditions for extraterrestrial forms of life [1,2]. One of the main goals of Mars investigation is "to follow the water" [3] as a prerequisite for the survival of living entities. Because of low temperatures (e.g., 215 K to 273 K at the equator) [3] and low Martian atmospheric pressure (600–800 Pa near the surface) [1], water can exist only as vapor, as ice, in brines [4], or bounded on the surface of the regolith as interfacial water in a liquid-like state [5], and it might also form by the process of deliquescence [6]. The water vapor is of particular interest, because it influences chemical reactions; because of the water content in the lower atmosphere and upper regolith, the phenomena like fog and thin frost layers occur, and could also be important for potential life forms. The question of water vapor content in the Martian atmosphere, with a dominant part of 96% carbon dioxide [7], can be resolved by accurate measurements of the relative humidity ($U_{w,i}$) in carbon dioxide at low temperatures and pressures. It is important to determine the metrological properties of humidity sensors in this extraterrestrial environment, prior to mounting onto a lander or rover mission or to use in Mars-simulating chambers.

This paper presents the results of the investigation on the SHT75 relative humidity sensors (Sensirion, Steafa ZH, Switzerland) in CO_2, performed at the DLR (German Aerospace Center) laboratory in Berlin. A similar investigation in the regular air of the Earth, performed in the same laboratory, has already been described in the literature [8]. That former paper [6] is essential for the understanding of the present paper, because the experimental setup, definitions of parameters, and so on, that are used in the present paper, have been explained and defined in the literature [8].

The SHT75 sensor is of polymer-based capacitive type, and both in the Mars simulation chambers on Earth and on the Martian rovers, that type of sensor is generally applied. For example, in the MESCH chamber (Mars Environmental Simulation Chamber) developed at the University of Aarhus (Aarhus C, Denmark), the Honeywell sensor HIH-3602C (Honeywell International Inc., Golden Valley, MN, USA) was used [9], and in the MARTE chamber (Mars environmental simulation chamber) built at Centro de Astrobiologia in Madrid, the Honeywell sensor HIH-4000 (Honeywell International Inc., Golden Valley, MN, USA) was employed [10]. In the PELS (Planetary Environmental Liquid Simulator) system at the University of Edinburgh, the Honeywell sensor HIH-4602-A (Honeywell International Inc., Golden Valley, MN, USA) was applied [11]. In the Phoenix spacecraft that landed on Mars in 2008, its instrument, MECA (Microscopy, Electrochemistry, and Conductivity Analyzer), contained a probe TECP (Thermal and Electrical Conductivity Probe) with the Panametrics sensor MiniCap-2 (GE Panametrics, Waltham, MA, USA) [12]. Most recently, the Curiosity rover, operating on Mars from 2012, contains the REMS instruments (Rover Environmental Monitoring Station) set with three Humicap sensors (Vaisala, Helsinki, Finland) [13]. The choice of polymer-based capacitance sensors is based on a number of advantages such as small dimensions and lightweight, low energy consumption, simple electronics for sensor's signal conditioning, a reliable measurement principle providing linear characteristics, and a relatively short response time. Sensirion AG belongs to the worldwide market-leading manufacturers of the humidity sensors, and the SHT75 sensor design exhibits high metrological properties.

2. Experimental Procedure

In the case of measurements in carbon dioxide, the number of measuring points (each point collected at stable pressure, temperature, humidity, and stable output signals of the reference dew point hygrometer and the investigated SHT75 sensors) taken into account (1316) was considerably smaller than for the measurements in air (5244), carried out in similar experiments at the DLR in 2013 [8]. Seven of the nine SHT75 sensors used in the air experiments were afterward tested in the carbon dioxide experiments.

The developed gas mixing system can generate either humidified air containing the amounts of water vapor that correspond to the humidity levels occurring in the atmospheric air on Earth, or the gas compositions corresponding to the atmosphere at the surface of Mars. The essential parts of the system are the adjustable mass flow controllers. As up to six individual gas components can be blended in the system, one controller per each inlet (including the control of the gas stream to be humidified) is used, and one controller per each of three outlets leading to the three measuring cells; nine controllers in total. In each measuring cell, three SHT75 sensors are placed. At 1013.25 hPa, the system can generate dew/frost points ranging from $-82\,°C$ dry air frost point (t_f) to $5\,°C$ dew point (t_d).

In place of the dry air that is used in the investigation described in the literature [8], in the experiments discussed here, CO_2 gas delivered in bottles was used as the carrier gas, with the purity of 99.995% and the frost point of $-66\,°C$. The system can provide a continuous flow of a carrier gas up to 150 L/h at $20\,°C$ and 1000 hPa [14]. A part of the dry carrier gas stream is saturated with water vapor when bubbled through liquid water in the scrubber bottles placed in the thermostat regulated water bath. That method ensures that a stable frost point temperature t_f of the humidified carrier gas with the setting accuracy of $\pm 0.5\,°C$ is maintained. When decompressing the humidified gas inside the measuring cells to 10 hPa with a vacuum pump, the dew point range of -94 to $-46\,°C$ can be reached. Only the results obtained from seven of the nine investigated sensors were considered to be valid for analyzing, because at the end of the experiments, the two sensors excluded from the analysis exhibited excessive deviations at the relative humidity above 80%.

The measurements in carbon dioxide were made in the temperature range of $10\,°C$ to $-70\,°C$ in 10 K steps, in monotonically decreasing order (nine temperature steps altogether). In each temperature step, the pressure was monotonically decreased. Firstly, the measurements at 1000 hPa were performed, and then the pressure was decreased down to 500 hPa, 200 hPa, and (from $-30\,°C$ downwards) to 10 hPa. Within every pressure step, firstly, the humidity was decreased in steps, and then increased back. For any combination of temperature and pressure values (i.e., of row and column headings of Table 1), the corresponding set of measurement points was taken only once.

In Table 1, the applied ranges of the relative humidity for every temperature–pressure combination are listed. The measurement points for 10 hPa and 200 hPa at $10\,°C$ and for 10 hPa at $0\,°C$, $-10\,°C$, and $-20\,°C$ could not be obtained because of the limitation of the gas mixing system. The measurement points at $-70\,°C$ for 500 hPa and 1000 hPa were not taken as a result of very long response times of the SHT75 sensors (tens of hours at higher pressure values).

Table 1. Range of the relative humidity of CO_2 under investigation (minimum and maximum values) for different temperature and pressure conditions. The relative humidity (Equation (1) and Section 3.2 in [8]) is calculated with respect to water U_w or ice U_i (marked by brackets). T denotes humid gas temperature (in °C).

T	p			
	1000 hPa	500 hPa	200 hPa	10 hPa
10 °C	84	37	-	-
	12	7	-	-
0 °C	70	74	30	-
	7	7	7	-
−10 °C	67 (74)	68 (75)	64 (71)	-
	21 (23)	5 (6)	6 (7)	-
−20 °C	65 (79)	67 (81)	70 (85)	-
	9 (11)	7 (9)	5 (7)	-
−30 °C	59 (80)	62 (83)	67 (90)	18 (24)
	16 (22)	4 (5)	6 (8)	8 (10)
−40 °C	54 (79)	53 (79)	56 (83)	56 (82)
	2 (3)	3 (5)	7 (10)	7 (10)
−50 °C	(100)	(97)	(87)	(100)
	(20)	(19)	(17)	(14)
−60 °C	-	(88)	(86)	(99)
	-	(18)	(19)	(14)
−70 °C	-	-	(96)	(91)
	-	-	(34)	(9)

3. Results

3.1. Pressure Dependency of the SHT75 in CO_2

Figure 1a–i show the pressure dependency of all of the SHT75 sensors investigated in CO_2 at various temperatures. In Figure 1e–g, each of the results of the fits at four different pressures are plotted. The reasons for the lack of one (Figure 1b–d,h) or two (Figure 1a,i) fit lines are explained above (comment on Table 1), or the slope of the fit line was too steep so that the resolution of the humidity readout strongly decreased, and an analysis has made no sense.

The pressure has a significant influence on the measured values of humidity, especially at lower pressure ranges. The slopes at 1000 hPa are always the greatest ones, whereas at 10 hPa (or if absent, at 200 hPa), the fitted lines have the least steep slopes. This pressure influence becomes more conspicuous with the temperature falling. At −30 °C, the value for the slope of the 1000 hPa fit line is only three times greater than that of the 10 hPa line, while at −50 °C the ratio is four to one.

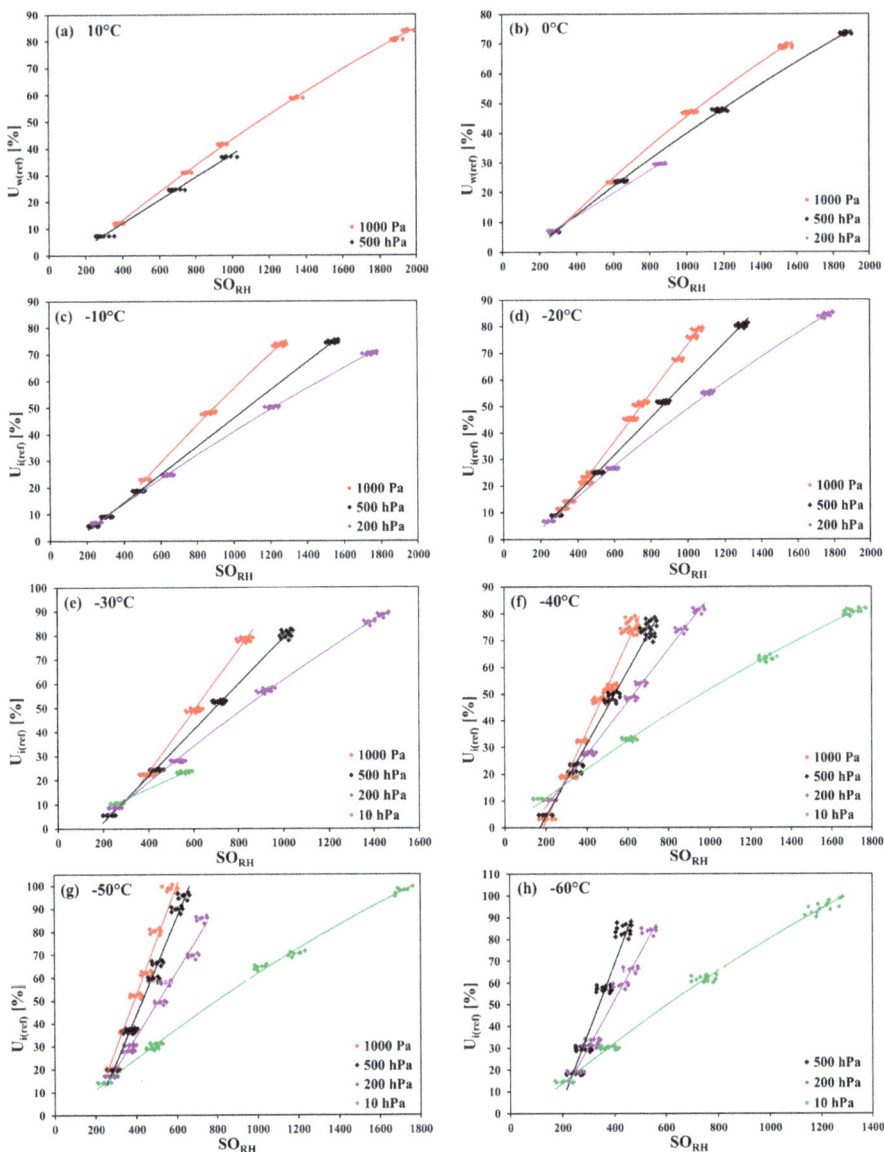

Figure 1. *Cont.*

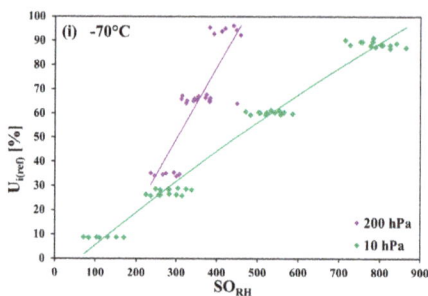

Figure 1. Pressure dependencies of the SHT75 sensors in CO_2 (regression lines fitted to measurement points collected from all tested sensors at different temperature/pressure combinations) at temperatures from $-70\,°C$ to $10\,°C$; SO_{RH} are the integer rough values (SO means 'sensor output', i.e., the humidity readout) of the SHT75 sensors.

3.2. Temperature Dependency of the SHT75 in CO_2

The set of fits depicted in Figure 1a–i (a separate figure for each constant temperature value) can be rearranged and divided into four other figures (a separate figure for each constant pressure value). Figure 2a–d show the temperature dependency of the fits at different pressures.

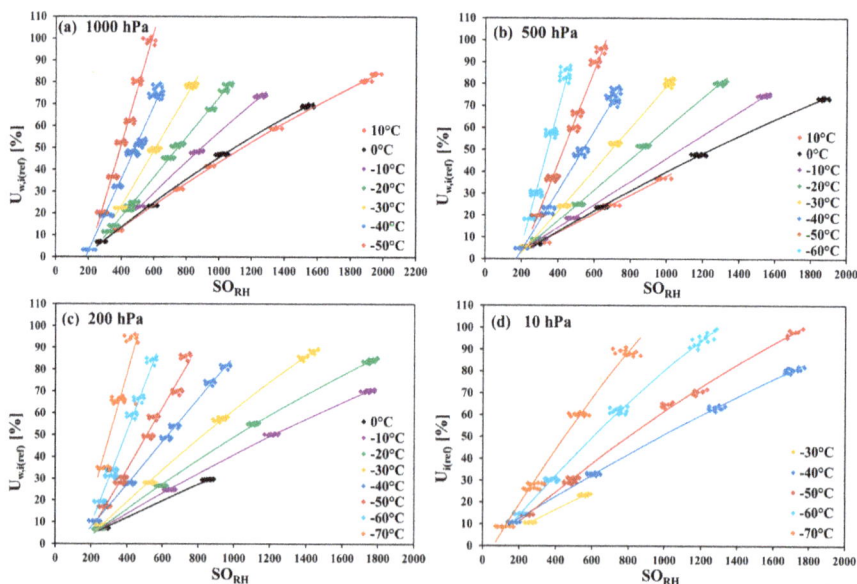

Figure 2. Temperature dependencies of the SHT75 sensors in CO_2 (regression lines fitted to measurement points collected from all tested sensors at different temperature-pressure combinations) at pressures from 10 hPa to 1000 hPa.

The four figures above show that the slope of the fit becomes greater with the decreasing temperature. The slope values at $10\,°C$ and $0\,°C$ are similar for 500 hPa and 1000 hPa. The difference increases at $-10\,°C$ and grows with decreasing temperature. For 1000 hPa, the slope value at $-50\,°C$ is four times greater than that at $10\,°C$. On the other hand, for 10 hPa, the slope value at $-70\,°C$ is only ca. three times greater than at $-30\,°C$. The fit equations are listed in Table 2.

Table 2. Polynomial (quadratic) and linear regression fit equations for each pressure–temperature pair shown in Figures 1a–i and 2a–d.

Pressure [hPa]	Temperature [°C]	Fit Equation
1000	10	$U_{w,i(ref)} = -0.0000053 \times SO_{RH}^2 + 0.0575737 \times SO_{RH} - 8.96$
	0	$U_{w,i(ref)} = -0.0000083 \times SO_{RH}^2 + 0.0648922 \times SO_{RH} - 11.04$
	−10	$U_{w,i(ref)} = -0.000008 \times SO_{RH}^2 + 0.0821515 \times SO_{RH} - 17.08$
	−20	$U_{w,i(ref)} = 0.0908286 \times SO_{RH} - 17.84$
	−30	$U_{w,i(ref)} = 0.1263865 \times SO_{RH} - 27.19$
	−40	$U_{w,i(ref)} = 0.1668640 \times SO_{RH} - 30.88$
	−50	$U_{w,i(ref)} = 0.2595892 \times SO_{RH} - 52.28$
500	10	$U_{w,i(ref)} = 0.0428904 \times SO_{RH} - 5.033$
	0	$U_{w,i(ref)} = -0.0000049 \times SO_{RH}^2 + 0.0523972 \times SO_{RH} - 7.62$
	−10	$U_{w,i(ref)} = 0.00000025 \times SO_{RH}^2 + 0.05227031 \times SO_{RH} - 6.55$
	−20	$U_{w,i(ref)} = 0.0703082 \times SO_{RH} - 10.59$
	−30	$U_{w,i(ref)} = 0.0952314 \times SO_{RH} - 16.2$
	−40	$U_{w,i(ref)} = 0.1362861 \times SO_{RH} - 23.27$
	−50	$U_{w,i(ref)} = 0.212559 \times SO_{RH} - 41.17$
	−60	$U_{w,i(ref)} = 0.3125689 \times SO_{RH} - 56.36$
200	0	$U_{w,i(ref)} = 0.0381205 \times SO_{RH} - 3.13$
	−10	$U_{w,i(ref)} = -0.0000043 \times SO_{RH}^2 + 0.0510756 \times SO_{RH} - 5.88$
	−20	$U_{w,i(ref)} = -0.0000067 \times SO_{RH}^2 + 0.064592 \times SO_{RH} - 8.87$
	−30	$U_{w,i(ref)} = -0.0000082 \times SO_{RH}^2 + 0.0817805 \times SO_{RH} - 12.12$
	−40	$U_{w,i(ref)} = 0.0978043 \times SO_{RH} - 11.71$
	−50	$U_{w,i(ref)} = 0.1414872 \times SO_{RH} - 22.15$
	−60	$U_{w,i(ref)} = 0.2155412 \times SO_{RH} - 34.06$
	−70	$U_{w,i(ref)} = 0.2965787 \times SO_{RH} - 39.6$
10	−30	$U_{w,i(ref)} = 0.0417898 \times SO_{RH} - 0.11$
	−40	$U_{w,i(ref)} = -0.0000045 \times SO_{RH}^2 + 0.0537153 \times SO_{RH} + 1.88$
	−50	$U_{w,i(ref)} = -0.0000086 \times SO_{RH}^2 + 0.0737212 \times SO_{RH} - 3.41$
	−60	$U_{w,i(ref)} = -0.0000152 \times SO_{RH}^2 + 0.1013665 \times SO_{RH} - 5.73$
	−70	$U_{w,i(ref)} = -0.000024 \times SO_{RH}^2 + 0.1409922 \times SO_{RH} - 8.25$

The equations in Table 2 are written in the following form:

$$U_{w,i(ref)} = a_2 \times SO_{RH}^2 + a_1 \times SO_{RH} + a_0 \text{ for polynomial (quadratic) regression fits,} \qquad (1)$$

$$U_{w,i(ref)} = a_1 \times SO_{RH} + a_0 \text{ for linear regression fits.} \qquad (2)$$

For each investigated pressure value, an individual regression equation is needed with its own slope and intercept value.

For exemplification, the relationship between temperature and the parameters a_0, a_1, and a_2 from Equations (1) and (2), is plotted for the pressures 1000 hPa and 10 hPa in Figure 3a–c (on the basis of the values from Table 2).

The polynomial regression fit equations of the temperature dependencies of the parameters a_0, a_1, and a_2 are listed in Table 3 (t denotes the sensor temperature).

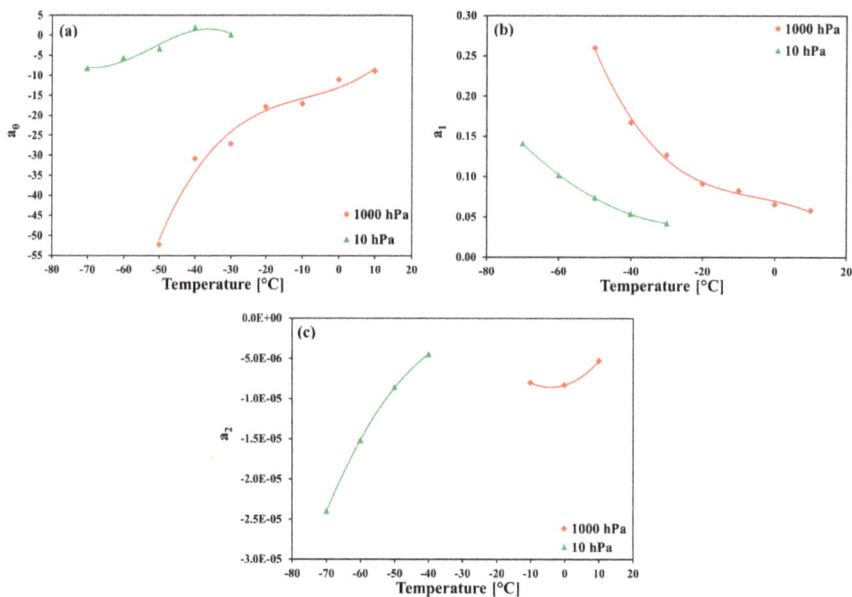

Figure 3. Temperature dependencies of the parameters a_0, a_1, and a_2 at pressures 1000 hPa and 10 hPa (on the basis of the values from Table 2).

Table 3. Polynomial regression fit equations of the curves shown in Figure 3a–c.

Pressure	Equation	Range of Validity
1000 hPa	$a_0 = 3.7153 \times 10^{-4}t^3 + 0.01015t^2 + 0.34192t - 13$	$-50\,°C$ to $10\,°C$
	$a_1 = -1.5502 \times 10^{-6}t^3 - 2.1957 \times 10^{-5}t^2 - 9.836 \times 10^{-4}t + 0.07$	$-50\,°C$ to $10\,°C$
	$a_2 = 1.65 \times 10^{-8}t^2 + 1.35 \times 10^{-7}t - 8.3 \times 10^{-6}$	$-10\,°C$ to $10\,°C$
10 hPa	$a_0 = -5.7126 \times 10^{-4}t^3 - 0.0897t^2 - 4.2482t - 61.84$	$-70\,°C$ to $-30\,°C$
	$a_1 = 4.5028 \times 10^{-5}t^2 + 2.0423 \times 10^{-3}t + 0.06285$	$-70\,°C$ to $-30\,°C$
	$a_2 = -1.175 \times 10^{-8}t^2 - 6.415 \times 10^{-7}t - 1.1345 \times 10^{-5}$	$-70\,°C$ to $-40\,°C$

3.3. Cross-Sensitivity of the SHT75 to CO$_2$

The substitution of the air with CO$_2$ as the carrier gas has led to some serious and unanticipated consequences. In exemplary Figure 4a–h, the characteristic curves based on the relative humidity values measured in the CO$_2$ atmosphere are compared with the curves based on the values measured in the air (taken from [8]), for the same selected temperature–pressure combinations. A conspicuous cross-sensitivity for any pressure and temperature is observed, with a strong increase at significantly decreased temperatures. For a given humidity value, this cross-sensitivity results in lowered SO_{RH} values being obtained in CO$_2$ and also in lower resolution, and a higher uncertainty of the measured values when compared with those obtained in the air.

In Figure 4a–h, eight pairs of linear (or slightly quadratic) characteristics are shown. Each pair consists of one line for a set of measurement points collected with the SHT75 sensors in air, and one line obtained in CO$_2$. The first related pair of figures (Figure 4a,b) are collated for 1000 hPa, and the following pairs are below—for 500 hPa, 200 hPa, and 10 hPa (Figure 4g,h). In each pair of figures, the fits at higher and lower temperature (for the same pressure value) are compared.

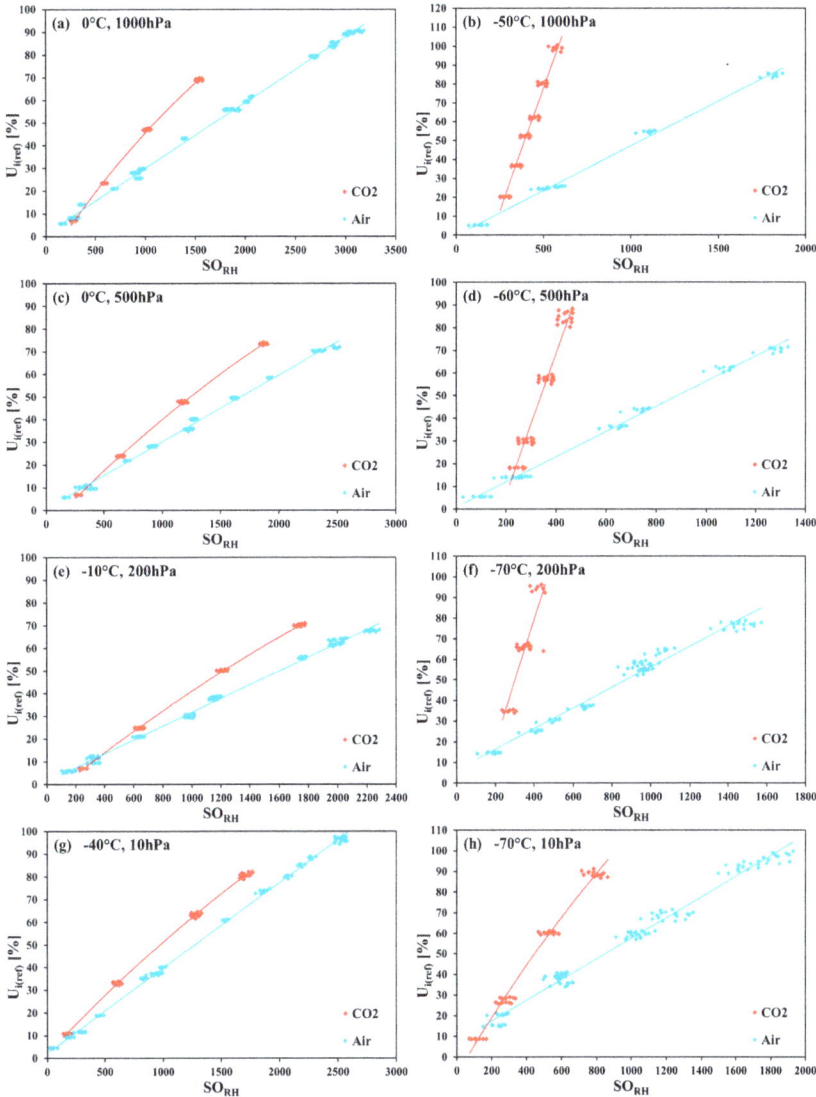

Figure 4. Exemplary collations of the SHT75 sensor characteristic curves (regression lines fitted to measurement points collected from all tested sensors) for the relative humidity measurement in CO_2 and in air, at selected temperature–pressure combinations.

In order to evaluate the strength of the cross-sensitivity, the ratio of two slopes, namely the slope of the linear characteristics obtained in CO_2 and in air (p and T being equal), can be used. That ratio for the right-column plots (low T) is for pressures of 200 hPa, 500 hPa, and 1000 hPa are 4.7, 4, and 3 times greater, respectively, than for the coupled left-column plots ($T = 0\,°C$ or $-10\,°C$). For $p = 10$ hPa, the ratio of the slopes calculated at and $-70\,°C$ is only 1.9 times greater than at $-40\,°C$. The comparison of Figure 4a–f (200 hPa to 1000 hPa) with Figure 4g–h (10 hPa) shows that the results of the measurements in a rarified carbon dioxide atmosphere are the most close to the measurements in air, but the best resolution is obtained at near-zero degrees Celsius temperatures.

4. Discussion

The measurements seem to prove a strong cross-sensitivity of the SHT75 sensors to CO_2. What could be the reason?

Inside the measuring system (Figure 1 in [8]), three main areas of the influence of carbon dioxide on the results may be suggested.

The first area of the possible interactions of carbon dioxide and water vapor is the gas mixing system, in which carbon dioxide is humidified and then mixed with dry carbon oxide in a given proportion, which determines the relative humidity of the mixture. During the mixing process, there is a possibility that some molecules of the gaseous carbon dioxide and the water vapor react to form molecules of gaseous carbonic acid (H_2CO_3). Some researchers suppose that the gaseous carbonic acid is present in cirrus clouds in the Earth's atmosphere and in the atmosphere on Mars [15]. In laboratory experiments, solid or gaseous carbonic acid is formed by the high-energy irradiation of H_2O/CO_2-ice or by acid–base chemistry at cryotemperatures. In the Earth's troposphere, under low humidity and at 250 K, the slow decomposition of H_2CO_3 is suggested. Under ambient conditions, the molecules of carbonic acid are very unstable if contact with water molecules is possible. However, a small portion of water vapor molecules might react with carbon dioxide, thereby reducing the amount of moisture measured by the sensor. In liquid water saturated with gaseous carbon dioxide, only ca. 0.2% of CO_2 [16] is bonded as carbonic acid; a similar proportion might be assumed in a mixture of water vapor and gaseous CO_2.

The second area is the chilled mirror surface of the dew point hygrometer. CO_2 or its reaction products could influence the dew point measurement. On the chilled mirror surface of the hygrometer, condensed water droplets or the deposition of frost crystals occur. The gaseous carbon dioxide is easily soluble in liquid water, especially at low temperatures, and also in the water droplets on the mirror. Then, inside the droplets, carbonic acid might form. But again, the molecules of carbonic acid inside the water droplets would be very unstable, and the droplets themselves evaporate frequently. In the case of frost, as the carbon dioxide molecules are large in comparison with water molecules in ice, the difference in kinetic diameters (H_2O [0.265 nm] vs. CO_2 [0.330 nm]) is unfavorable for the penetration of CO_2 into frost crystals. CO_2 could also freeze out on the chilled mirror surface, but the sublimation point is −78.5 °C at 1013.25 hPa [17] and the hygrometer has not reached such low temperature at 1000 hPa. Finally, the calculated values from the gas mixing system were in agreement with the values measured with the hygrometer. For these reasons, the use of a dew point mirror hygrometer should not cause the strong deviation of the SHT75 sensor measurement values in CO_2 from the values measured in air.

The third area is the sensing layer of the sensor. A competition of water vapor molecules and CO_2 molecules for the access to functional groups of an adsorbing surface was observed in the case of carboxyl groups on the carbon surface [18]. The polarity of water molecules is well known, but also, the carbon dioxide molecules exhibit a slight polarity. In the O=C=O molecule considered as quasi-linear, the end oxygen atoms are slightly electronegative, whereas the slight positive charge is located near the central carbon atom [19]. The polymers applied as a sensing layer in the capacitive humidity sensors are mostly polyimide-based, and the most popular are the various polyimides similar to Kapton®. In the Kapton® structure, the carboxyl groups –C=O are the primary bonding sites for the adsorption of water vapor molecules. Also, the ether groups C–O–C and N–C groups can constitute adsorption sites [20].

The most probable explanation for a strong cross-sensitivity of SHT75 humidity sensors to carbon dioxide assumes that the molecules of CO_2 interact with the adsorption sites on the thin polymer layer. Firstly, the CO_2 molecules can produce weak hydrogen bonds between the O or N atoms in the polymer chains, and the carbon atom in the CO_2 molecule. Secondly, the CO_2 molecules can attach to water molecules that have created hydrogen bondings at primary adsorption sites. Then, no more water molecules can be adsorbed as dimers or clusters at an adsorption site blocked by CO_2 molecules, and the amount of moisture adsorbed on the polymer sensing layer is strongly reduced.

This explanation may justify the steepest slopes (and the smallest output signals) at higher pressures of the humidified CO_2. The increase of the slope values when the temperature falls, although observed in air, is much stronger in carbon dioxide. Here, the explanation could be the difficulties in penetrating the water molecules inside the polymer layer when the thermal movements of the polymer chains are reduced, together with the presence of relatively bigger carbon dioxide molecules adsorbed on the polymer sites, which partly block the ways of penetration for much smaller water vapor molecules.

Based on the collected data, this kind of sensor seems to be a reasonable choice for application in the harsh environment on Mars. A major disadvantage is the limitation of the measurement range, down to about $U_i = 5\%$ for low humidity. That limit could be reached when the temperature of the atmosphere surrounding the sensor is above $-55\ °C$ at a frost point of $t_{fp} = -76\ °C$ [21]. Thus, for the measurement of the U_i or U_w values below that lower range limit of the polymer-based capacitive sensors, at higher temperatures, another sensor working principle is necessary (e.g., that of a coulometric sensor). The high U_i values above 95% can also be difficult to measure. For the measurement of the frost point ($U_i = 100\%$), a thin plate coupled with a precise temperature measurement that allows for detecting the adsorption and desorption of condensed water on the plate, could be used. Such considerations and measurements are described in the literature [22].

5. Conclusions

The measurements of the relative humidity of the gaseous carbon dioxide using polymer-based capacitive humidity sensors revealed a strong dependence of the sensor characteristic curve on both the temperature and the pressure of the measured humid gas. The greatest deviation from the sensor nominal characteristic curve was observed at the lowest investigated temperature of $-70\ °C$ and at the pressures 200 hPa, 500 hPa, and 1000 hPa.

The comparison with the results obtained for the same sensors in the measurements of humid air showed big discrepancies that demonstrate the considerable cross-sensitivity of the sensors to carbon dioxide. The most probable explanation of this effect can be the interactions of carbon dioxide molecules both directly with the adsorption sites on the polymer layer, and indirectly with the water molecules adsorbed on the primary adsorption sites on the polymer.

Despite of the observed cross-sensitivity, the polymer-based capacitive sensors can still be used for the measurements of relative humidity in carbon dioxide at low pressures, within a broad range of temperatures. An individual calibration of the sensors for such applications is recommended, and earlier experiments should be checked for whether the cross-sensitivity has not been taken into account. The research on cross-sensitivity to various gases for this type of humidity sensor should be continued.

Author Contributions: A.L. conceived the methodology, and designed and performed the experiments. J.M. analyzed the data and prepared the manuscript.

Acknowledgments: Funding from the German Federal Ministry of Economics and Technology-BMWi (project No. SF11021A) is gratefully acknowledged.

References

1. Rummel, J.D.; Beaty, D.W.; Jones, M.A.; Bakermans, C.; Barlow, N.G.; Boston, P.J.; Chevrier, V.F.; Clark, B.C.; de Vera, J.-P.P.; Gough, R.V.; et al. A New Analysis of Mars "Special Regions": Findings of the Second MEPAG Special Regions Science Analysis Group (SR-SAG2). *Astrobiology* **2014**, *14*, 887–968. [CrossRef] [PubMed]
2. Domagal-Goldman, S.D.; Wright, K.E.; Adamala, K.; Arina de la Rubia, L.; Bond, J.; Dartnell, L.R.; Goldman, A.D.; Lynch, K.; Naud, M.-E.; Paulino-Lima, I.G.; et al. The Astrobiology Primer v2.0. *Astrobiology* **2016**, *16*, 561–653. [CrossRef] [PubMed]

3. National Research Council; Division on Earth and Life Studies; Board on Life Sciences; Division on Engineering and Physical Sciences; Space Studies Board. *Committee on an Astrobiology Strategy for the Exploration of Mars. An Astrobiology Strategy for the Exploration of Mars*; National Academies Press: Washington, DC, USA, 2007; ISBN 978-0-309-10851-5.

4. Kereszturi, A.; Möhlmann, D.; Berczi, S.; Ganti, T.; Horvath, A.; Kuti, A.; Sik, A.; Szathmary, E. Indications of brine related local seepage phenomena on the northern hemisphere of Mars. *Icarus* **2010**, *207*, 149–164. [CrossRef]

5. Lorek, A.; Wagner, N. Supercooled interfacial water in fine-grained soils probed by dielectric spectroscopy. *Cryosphere* **2013**, *7*, 1839–1855. [CrossRef]

6. Pál, B.; Kereszturi, Á. Possibility of microscopic liquid water formation at landing sites on Mars and their observational potential. *Icarus* **2017**, *282*, 84–92. [CrossRef]

7. Mahaffy, P.R.; Webster, C.R.; Atreya, S.K.; Franz, H.; Wong, M.; Conrad, P.G.; Harpold, D.; Jones, J.J.; Leshin, L.A.; Manning, H.; et al. Abundance and Isotopic Composition of Gases in the Martian Atmosphere from the Curiosity Rover. *Science* **2013**, *341*, 263–266. [CrossRef] [PubMed]

8. Lorek, A. Humidity measurement with capacitive humidity sensors between −70 °C and 25 °C in low vacuum. *J. Sens. Sens. Syst.* **2014**, *3*, 177–185. [CrossRef]

9. Jensen, L.L.; Merrison, J.; Hansen, A.A.; Mikkelsen, K.A.; Kristoffersen, T.; Nørnberg, P.; Lomstein, B.A.; Finster, K. A Facility for Long-Term Mars Simulation Experiments: The Mars Environmental Simulation Chamber (MESCH). *Astrobiology* **2008**, *8*, 537–548. [CrossRef] [PubMed]

10. Sobrado, J.M.; Martín-Soler, J.; Martín-Gago, J.A. Mimicking Mars: A vacuum simulation chamber for testing environmental instrumentation for Mars exploration. *Rev. Sci. Instrum.* **2014**, *85*, 035111. [CrossRef] [PubMed]

11. Martin, D.; Cockell, C.S. PELS (Planetary Environmental Liquid Simulator): A New Type of Simulation Facility to Study Extraterrestrial Aqueous Environments. *Astrobiology* **2015**, *15*, 111–118. [CrossRef] [PubMed]

12. Zent, A.P.; Hecht, M.H.; Hudson, T.L.; Wood, S.E.; Chevrier, V.F. A revised calibration function and results for the Phoenix mission TECP relative humidity sensor: Phoenix Humidity Results. *J. Geophys. Res. Planets* **2016**, *121*, 626–651. [CrossRef]

13. Gómez-Elvira, J.; Armiens, C.; Castañer, L.; Domínguez, M.; Genzer, M.; Gómez, F.; Haberle, R.; Harri, A.-M.; Jiménez, V.; Kahanpää, H.; et al. REMS: The Environmental Sensor Suite for the Mars Science Laboratory Rover. *Space Sci. Rev.* **2012**, *170*, 583–640. [CrossRef]

14. Lorek, A.; Koncz, A. Simulation and measurement of extraterrestrial conditions for experiments on habitability with respect to Mars. In *Habitability of Other Planets and Satellites*; de Vera, J.-P., Seckbach, J., Eds.; Springer: Dordrecht, The Netherlands, 2013; Volume 28, pp. 145–162. ISBN 978-94-007-6545-0.

15. Ghoshal, S.; Hazra, M.K. $H_2CO_3 \rightarrow CO_2 + H_2O$ decomposition in the presence of H_2O, HCOOH, CH_3COOH, H_2SO_4 and HO_2 radical: Instability of the gas-phase H_2CO_3 molecule in the troposphere and lower stratosphere. *RSC Adv.* **2015**, *5*, 17623–17635. [CrossRef]

16. Harris, D.C. *Lehrbuch der Quantitativen Analyse*; Springer-Verlag: Berlin, Germany, 2014; ISBN 978-3-642-37787-7.

17. Hartmann-Schreier, J. Kohlendioxid. RÖMPP Online. Available online: https://roempp.thieme.de/roempp4.0/do/data/RD-11-01458 (accessed on 28 February 2018).

18. Nishino, J. Adsorption of water vapor and carbon dioxide at carboxylic functional groups on the surface of coal. *Fuel* **2001**, *80*, 757–764. [CrossRef]

19. Xing, W.; Liu, C.; Zhou, Z.; Zhou, J.; Wang, G.; Zhuo, S.; Xue, Q.; Song, L.; Yan, Z. Oxygen-containing functional group-facilitated CO_2 capture by carbide-derived carbons. *Nanoscale Res. Lett.* **2014**, *9*, 189. [CrossRef] [PubMed]

20. Majewski, J. Low Humidity Characteristics of Polymer-Based Capacitive Humidity Sensors. *Metrol. Meas. Syst.* **2017**, *24*. [CrossRef]

21. Ryan, J.A.; Sharman, R.D. H_2O frost point detection on Mars? *J. Geophys. Res.* **1981**, *86*, 503. [CrossRef]

22. Koncz, A. Entwicklung und Schaffung eines in-situ Feuchtemessgerätes für den Mars im Zusammenhang mit der ESA Marsmission ExoMars. Ph.D. Thesis, Universität Stuttgart, Stuttgart, Germany, May 2012.

sensors

MDPI

Article

Online Moisture Measurement of Bio Fuel at a Paper Mill Employing a Microwave Resonator †

Martta-Kaisa Olkkonen

Department of Electrical Engineering and Automation, Aalto University, Maarintie 8, 02150 Espoo, Finland;
martta-kaisa.olkkonen@alumni.aalto.fi

† This paper is an extended version of my paper published in 9th International Conference on Electromagnetic Wave Interaction with Water and Moist Substances (ISEMA 2011), Kansas City, MO, USA, 31 May–3 June 2011.

Received: 1 October 2018; Accepted: 6 November 2018; Published: 9 November 2018

Abstract: This paper investigates online moisture measurement of biofuel employing a strip line cavity resonator at approximately 366 MHz, attached above and below the conveyor belt. An existing sensor design is modified for the factory assembly, and the correct operation has been tested prior to this paper with a small number of measurement points and collected reference samples ($n = 67$). The purpose is now to concentrate on the accuracy of the measurement and increase the number of measurement points ($n = 367$). The measurements were made in 5 different lots, and the thickness and moisture properties of the biomaterial mat were varied between minimum and maximum levels by adjusting the settings of the belt filter press that presses pulp slush into a mat. In order to further reduce inaccuracy, at the maximum one standard deviation was allowed from the average height of the equivalent water layer for each dataset, and consequently the number of samples was reduced to 235. A linear fit and a parabola fit were determined for thickness of the equivalent water layer vs. the relative resonant frequency shift: $R^2 = 0.82$ and $R^2 = 0.78$.

Keywords: bio fuel; microwave resonator; moisture measurement; paper mill

---

1. Introduction

1.1. Microwave Moisture Measurement of Biomaterials

Moisture content monitoring is a method for evaluating an industrial process; this paper investigates online moisture measurement of biofuel conducted in a paper factory. Moisture measurements of biomaterials such as paper are discussed in [1,2]. The biomaterial is slush of the paper process and it is pressed into a mat in a belt filter press used in biological treatment plants for processing pulp slush. It is characterized by a porous, inhomogeneous structure and also a high ion content due to impurities, mainly salts. For example [3] studied quality assessment of the paper manufacturing process. On-line moisture measurements are challenging, because the material under test moves constantly as it passes the sensor on a conveyor belt or in a pipe. Ref. [4] reports moisture content monitoring in paper and veneer manufacturing processes employing similar kinds of resonator sensors as in the current research. A resonator sensor was patented (2012) for measuring the moisture content of a mat in the wire end of a paper machine [5]. However, in our present application, an extra variable compared to plywood is thickness, because the belt filter press is not designed to adjust the thickness of the biomaterial accurately, as typically it is not a relevant parameter. This feasibility study on moisture evaluation of biofuel is especially challenging because it is conducted as an on-line measurement and the material is rather inhomogeneous. The mineral content of biomaterials can vary highly and finding a correlation between an electrical parameter and the water content requires a proper statistical analysis.

Analysis of biofuel drying was carried out in [6] for wood-based masses and in [7] a case study was carried out for tea leaves. Typical biofuels are bark, forest residue, sawdust and crushed construction wood [6]. The microwave moisture measurement methods are well-established for wood-base biofuels, since the measurement instrumentation of for example timber is already found in the literature [8]. The moisture content generally varies from 45–55% (wet-basis) in biofuel but this causes combustion problems in power plants [9]. The price of biofuel is determined by the moisture content and, in addition, water is not a part of the combustion process; instead it only absorbs heat energy during the burning. Typical methods for measuring the moisture content are either fuel flow online or fuel bulk in a large container [10].

One traditional and destructive moisture evaluation method is oven-drying, where a sample is taken from the conveyor belt and dried in the oven; for wood the standard temperature is 102–105 degrees Celsius. The percentage moisture content is calculated on a wet basis as a ratio of mass of water to the total mass of moist material (gravimetric)

$$M = \frac{m_w}{m_w + m_d} \times 100\%, \tag{1}$$

where m_w is the mass of water and m_d is the mass of dry material.

1.2. Definition of Electric Parameters

The complex permittivity of an isotropic material is

$$\varepsilon = \varepsilon_r \varepsilon_0 = \varepsilon_0(\varepsilon_r' - j\varepsilon_r''), \tag{2}$$

where ε_r' is the real part and ε_r'' is the imaginary part of the complex relative permittivity and the permittivity of vacuum is $\varepsilon_0 \approx 8.854 \times 10^{-12}$ F/m.

Let us define the parameters that are used for deriving the material properties from measurement with a strip line cavity resonator sensor. The perturbation equations [11] (pp. 141–145) assume that the electric field is approximately unchanged when the change in permittivity of material is small. The electric field needs to be constant inside the sample that is inserted in the resonator. Even mode occurs when the electric field is tangential to the sample. The following approximation for the relative change of frequency (3) applies for the even wave mode [11]:

$$\frac{\Delta f_r}{f_r} \approx -\frac{\varepsilon_r' - 1}{2} S, \tag{3}$$

when the relative magnetic permeability is $\mu_r \approx 1$ and S is the filling factor of the resonator and it depends on the dimensions of the sample. Equation (3) is essentially a linear approximation of the frequency dependency on the real part of relative permittivity, since it assumes a low loss case, so that $\varepsilon_r' \gg \varepsilon_r''$.

The moisture measurement is conducted as a single-parameter measurement according to (3) that is not dependent on ε_r''. The real part of permittivity is simply related to the resonant frequency shift caused by inserting dielectric material in the resonator cavity. As a further approximation, the only variable is limited to the changing water content in the material. Thus, we aim to find a statistical correlation directly between the moisture content and the resonant frequency shift of the resonator.

1.3. Objective of Research

The pulp slush is first treated by adding polymers, which make the slush lose water more easily. The moisture measurement is conducted online, but the material is also conveyed to a combustion chamber before burning. Therefore, an interesting result of this research is not the moisture content at different points of the mat but, instead, the total amount of water in the container. In other words, the objective of this work is not to find wet patches on the mat, but instead to estimate the average moisture content of the material that passes the sensor. Based on the moisture measurement, it is possible to estimate the amount of water in the container, but many samples need to be recorded so that the measurement would be representative of the bulk [9]. Since the microwave moisture measurement represents a volumetric method, the process can be monitored during a specified time and can estimate the amount of water of the bulk.

The requirement specifications for the project are as follows: for the biofuel plant, the goal would be to find an optimal level for the dehumidification of the biomaterial prior to burning. The primary objective of burning the biomaterial is to minimize its volume, which reduces the disposal costs. The heat energy is significant for the plant only if the dry material content in the biomaterial mat is sufficient. The dry material content, which is achieved by mechanical pressing, is usually between 20–40%, an optimum being 50% for the burning according to the biofuel plant. Thus, the moisture content needs to be monitored in order to find an optimum balance where the costs of burning the moist material are lower than the possible savings in energy. If the accuracy of the moisture measurement is sufficient, the paper factory can decide whether the biomaterial is burnt for energy or disposed of. Also, the process parameters of the belt filter press could be more easily adjusted according to the moisture readings. The main objective of this first experiment was to investigate, whether the chosen sensor type could be used for monitoring the moisture content of the biomaterial that differs from paper or wood. Even after pressing, the moisture content of the biomaterial can reach up to 60%, whereas for paper it stays usually under 10%. The biggest difference compared to paper is a large ionic impurity content, mainly salts, which cause conductive losses.

2. Materials and Methods

Online moisture measurement requires a non-contacting transmission-type resonator; Material flows through the sensor and the so-called transmission scattering parameters are measured and analyzed. Due to inhomogeneity of the material, a relatively large footprint of the order of 300 cm^2 is chosen in order to obtain an average result of the footprint area. This leads to choosing the frequency range of the measurement (~400 MHz) to be rather low. The chosen sensor type for this first feasibility study is an existing strip line cavity resonator design [12], which was modified for the factory assembly. This kind of transmission measurement averages a volume that can be called an equivalent volume element. This volume element is determined by the footprint area of the measurement and the thickness of the mat. In online sensing the speed of the measurement should be optimal so that the device does not move more than the area of the footprint during the recording of one measurement point. The permittivity value becomes a "sliding average" if the speed of the measurement device is too high compared to the measurement speed. However, this can be allowed according to the requirement specification of finding the volumetric water content of the material that is burnt.

2.1. Modification of the Existing Sensor Design and Simulations

The chosen sensor is a $\lambda/4$ strip line cavity resonator, and the field lines appear as in a strip transmission line. It was patented in [12] and in commercial use for measuring the moisture content of plywood [13]. A similar resonator was used for measurement of a paper web in [14]. The one-conductor strip line resonator supports only the odd mode and the two-conductor $\lambda/4$ strip line resonator supports only the even mode [14]. The odd and even modes are partly sensitive to different parameters of the material. For this resonator, the relative resonant frequency shift is derived from the perturbation theory [11].

The dimensions of the resonator halves are 250 mm × 300 mm × 100 mm and the strip is placed at 60 mm distance from the bottom of both resonator halves. It was not possible to assemble the existing rectangular resonator to the belt filter press and, therefore, as a modification to the original design, one edge of both resonator halves was made slanted. The dimensions of a resonator half in millimeters are shown from the side in Figure 1a and from the top in Figure 1b.

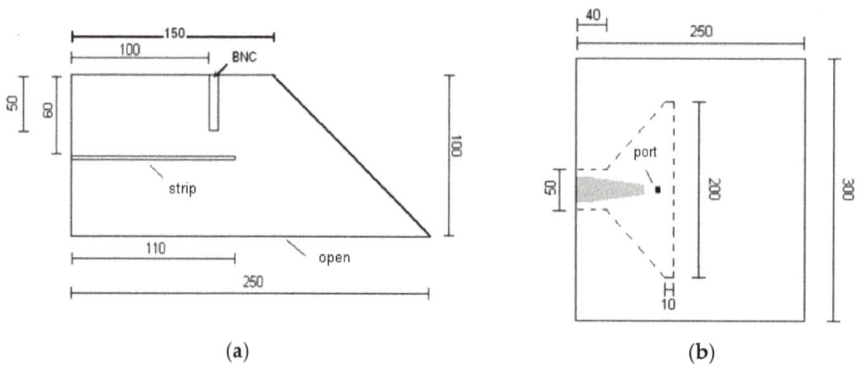

Figure 1. (a) Side view of a resonator half; (b) top view of a resonator half. Dimensions are in millimeters.

Operation of the resonator was simulated and the distance between the resonator halves was also set to 60 mm. According to the simulations, the slant edge had negligible effect on the resonator operation because the resonant frequency was 359.5 MHz for the rectangular resonator and 357.5 MHz with slant edges, respectively.

Preliminary tests with the modified sensor design were carried out in [15,16]. Figure 2a presents the resonator construction during laboratory tests, where the resonant frequency was measured without the material under test (MUT) and it was f_{r0} = 366.3 MHz. The vector network analyzer employed in the laboratory and also in the factory experiments was HP8753D. Figure 2b shows the resonant curve of the empty resonator; the insertion loss is 1.2 dB at the resonant frequency. The resonant frequency is determined accurately only from a sharp resonance. Insertion loss varied during the measurement campaign of the biofuel between 26 and 24 dB of those samples that showed a sharp detectable resonance; other samples were not included in the analysis.

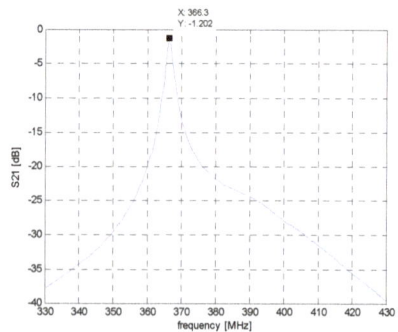

Figure 2. (a) Resonator in the laboratory; (b) measured resonance curve of the empty resonator.

2.2. Online Moisture Measurement of Biofuel

The sensor was assembled to the end part of a belt filter press, where the mat falls to a ripper, cutting the mat into small pieces and after that the material is moved to a combustion chamber. Figure 3 shows the lower resonator half and the center conductor or "strip". The strips of the resonator were made wider (200 mm) to enlarge the measurement footprint. A Plexiglas protects the cavity and directs the mat to flow evenly through the resonator halves.

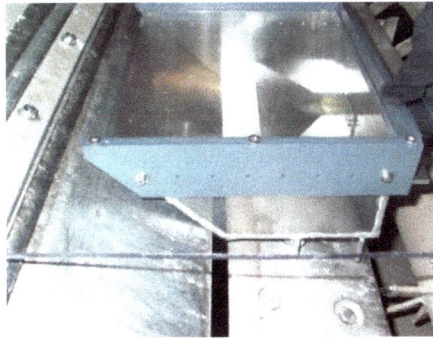

Figure 3. Resonator half assembled to the belt filter press.

Figure 4 illustrates how the biomaterial mat flows through the resonator halves in the online measurement. The sensor was placed in the middle of the mat and the measurement was made along one longitudinal line. The uncertainty of the relative resonant frequency shift was 1.3% when only 65 samples were analyzed [16]. Therefore, it was presumed that the number of samples should be increased to obtain better accuracy. For the new analysis, in total 367 measurements were made and the same number of reference samples was collected. The measurements were made in 5 different lots, and the thickness and moisture properties of the biomaterial mat were varied between the lots by adjusting the settings of the belt filter press differently for each of them.

Figure 4. Biomaterial mat moving through the resonator halves during the factory experiments.

The reference samples were collected from the mat after the corresponding resonant frequencies were recorded. The area and thickness of the samples were measured in the laboratory prior to oven-drying and they were on average 96 mm^2 and 14.3 mm, the average volume of the samples being 138 cm^3. The samples were oven-dried and weighed before and after in the laboratory for the gravimetric determination of the moisture content. For wood, the standard oven-drying temperature is typically 100–105 °C [17]. There is no standard for the biofuel in question but as a precaution, the temperature of the oven was set only to 80 °C so that volatilization of organic matter would not occur; the biomaterial consists of a high amount of microbes and some chemicals that are added before burning. The relative water content (wet-basis) of the reference samples determined using (1) varied between 47–67%.

3. Results

In an online microwave measurement, the sensor does not measure a gravimetric moisture content of discrete "samples", but instead it represents a volumetric measurement of a moving mat. The effective volume of the measurement would be defined as the thickness of the mat multiplied by the measurement footprint, which is determined by the width of the center conductor (200 mm) of the strip line cavity resonator. The measurement footprint is slightly larger than the area of the center strip, the half-power area being approximately 300 mm × 100 mm [14]. Therefore, the average area of the reference samples (96 mm^2) was clearly smaller than the footprint of the measurement. However, the accurate volume element of each measurement point cannot be defined because the thickness changes constantly and the sensor does not obtain information about the rate of this change.

Normally one would find a correlation between the relative water content and resonant frequency shift. However, in this current application, the thicker the material, the lower its relative water content, but the absolute water content is higher than for a thinner mat. This is related to the operation of the belt filter press, and it is necessary to model the water content as an absolute quantity such as thickness of the equivalent water layer [16]. The equivalent water layer is not tightly dependent on the thickness of the material. In principle, the same amount of dry material is inserted in the press constantly, and pressing affects the density and water content of the mat. If the belt filter press is adjusted to produce a thinner mat, this consequently results in a mat which is denser. Alternatively, a less tightly pressed mat is not as dense but contains more water.

The equivalent water layer corresponds to the amount of water per surface area in each sample, which essentially the sensor measures, considering the assumption of the negligible effect of the dry material. The height of the equivalent water layer is equivalent to the mass per area of water through the following equation:

$$h_{eq}[\text{mm}] = \frac{V_{water}}{A_{sample}} \Delta \frac{m_w[\text{g}]}{10^3[\text{g/m}^3]} \cdot \frac{1000}{A_{sample}[\text{m}^2]} = m_{aw}[\text{kg/m}^2]. \tag{4}$$

Finding a suitable correlation between the water content and ε'_r is a complicated task and depends on many parameters, e.g., on frequency; for pure water, the real part of permittivity is independent of frequency below the relaxation frequency.

Statistical correlation was determined between the measured parameter and the material parameter in [18], where the moisture content of single soy beans was measured using a resonator. The dependency between the resonant frequency shift and water content was found to be linear for a constant dry mass. Ref. [19] reported that the relation between the moisture content and ε'_r (at 18 GHz) of wool was non-linear but, above 20% of moisture, the permittivity increased more steeply, following a linear trend. This kind of behavior was attributed to the bound-water effect. As a first estimate, a linear relation is assumed also between the permittivity of the biomaterial mat and the water content. In addition, the resonator sensor type is designed to support the even mode and be linear as a function of ($\varepsilon'_r - 1$), based on the approximation of the perturbation formula [11]. The result of the moisture measurement is presented in Figure 5, with a linear fit. Figure 6 shows the corresponding residual plot of the linear fit.

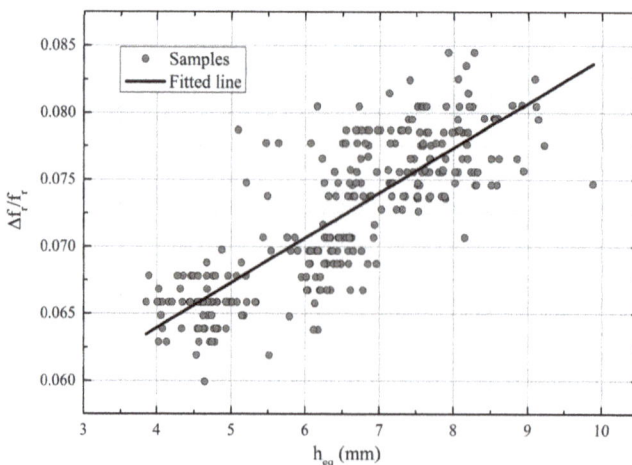

Figure 5. The relative resonant frequency shift vs. thickness of the equivalent water layer.

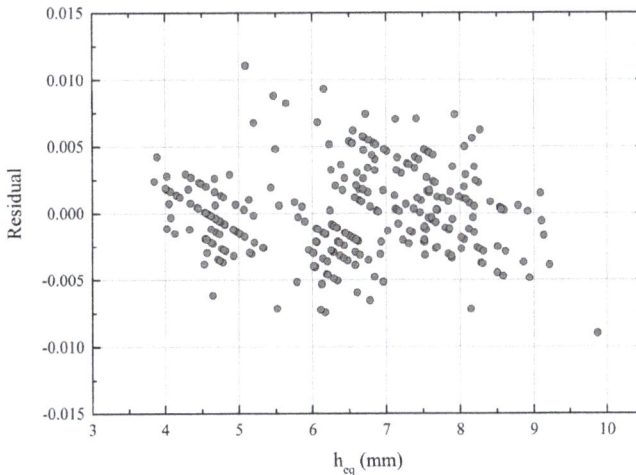

Figure 6. Residual plot for linear fit.

In an on-line measurement variations due to constant changes in thickness of the moving material pose a challenge. Feasibility of the single-parameter measurement is based on an estimation that the dry material has a significantly smaller effect on the resonant frequency than water. However, changes in the thickness or density also cause uncertainty in the measurement of the water content. The thicknesses of the samples were measured after collecting them from the mat, and they varied between 7 mm to 21.4 mm.

Another analysis of the results was made to diminish the uncertainty due to the large variations in the thickness of the material. The number of samples was reduced so that fewer water content levels share the same resonant frequency shift. For each measurement data set (5 in total), the average height of the water layer is different because of the settings in the filter press system. The average thickness of the mat with corresponding standard deviation (STD), and thickness of the equivalent water layer is listed for each series in Table 1. The range of resonant frequency shifts was determined so that at maximum one standard deviation was allowed from the average height of the equivalent water layer for each dataset and, consequently, the number of samples was reduced to 235. The measurement resolution of the resonant frequency was 0.35 MHz, because the number of points was 201 in the band 310 to 380 MHz. The thickness of the equivalent water layer varied now only approximately ±1 mm per 0.35 MHz or per 0.1 percentage points shift of the relative resonant frequency, which was essentially the "resolution" in Figure 5.

Table 1. Average thickness of the equivalent water layer with corresponding standard deviation (STD) and the average thickness of the mat for the measurement series I–V.

Measurement Series	I	II	III	IV	V
Average thickness of equivalent water layer (mm)	4.79	8.40	7.74	6.31	6.91
STD (mm)	0.49	0.59	0.63	0.39	0.64
Average thickness of the mat (mm)	9.2	20.5	16.0	10.5	15.4

Figure 7 presents the thickness of the equivalent water layer across the relative resonant frequency shift. Both a linear fit and a parabola fit were determined and they gave $R^2 = 0.78$, $R^2 = 0.82$, respectively. Figure 8 shows the residual of the linear and parabola fit and the corresponding parameters are listed in Table 2. Data sets I and IV are distinguishable but sets II, III and V lie partly on top of each other between water levels 7 mm and 9 mm. Between the water levels 8 mm and 9 mm, the resonant frequency shift is higher than would be expected based on the linear fit or parabola fit. Such a phenomenon was observed in [20], where wet paper was measured with a similar strip line resonator, even though the dependency was expected to be linear based on theory. Their assumption was that at higher values of ε'_r, determination of the moisture content, and thickness is more uncertain of very wet paper (>50%).

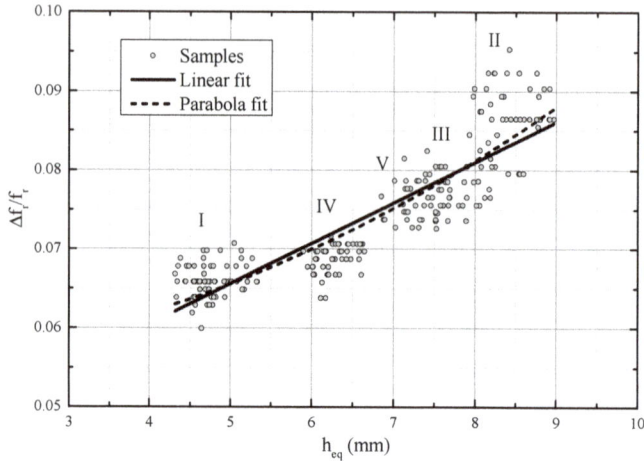

Figure 7. Thickness of the equivalent water layer vs. the relative resonant frequency shift with a reduced number of samples. A linear fit ($R^2 = 0.78$) and a parabola fit ($R^2 = 0.82$).

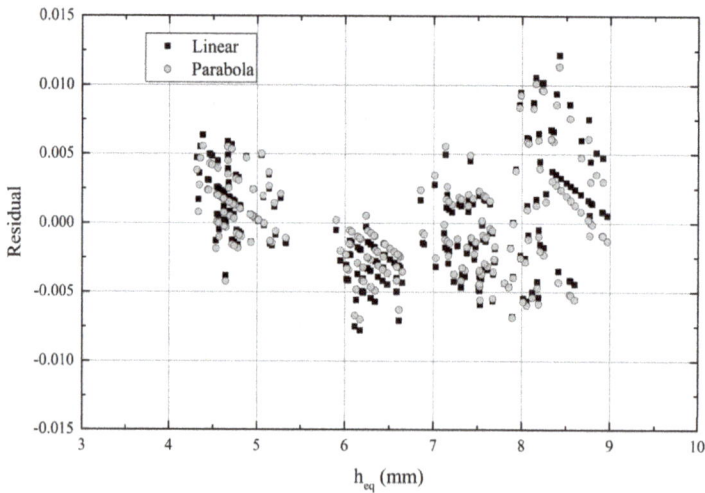

Figure 8. Residuals for parabola and linear fit.

Table 2. Parameters of the linear and parabola fit.

Equation	Linear (Orig. Data Figure 5)	Linear (Datasets I–V, Figure 7)	Parabola (Datasets I–V, Figure 7)
	$y = A + Bx$	$y = A + Bx$	$y = A + Bx + Cx^2$
A	0.0505	0.03998	0.05554
B	0.0034	0.00512	0
C	-	-	4.0073×10^{-4}
R^2	0.67	0.78	0.82
Residual min	−0.00898	−0.0078	−0.007
Residual max	0.01108	0.01218	0.01136

Table 2 lists the parameters for linear fit of original data (Figure 5), as well as the linear and parabola fit after reducing the number of data points according to Table 1.

4. Conclusions

4.1. Discussion

This paper carried out new measurements and calculations as part of a feasibility study on the moisture measurement of biofuel. The employed sensor type is an existing design which is in commercial use for moisture measurement of veneer. The research was conducted in collaboration with a Finnish paper factory. When the biomaterial is highly inhomogeneous in particular, point-like resonant frequency measurements are not informative about the water content of the material as such, but instead statistical processing and analysis of the data is required.

The operation of the belt filter press is not directly related to actual measured resonant frequency shifts. As a compromise for taking into account thickness variation of the samples and uneven distribution of the relative moisture content, the water content in each sample was modelled as an equivalent water layer.

The operation of the press is not designed so that the thickness could be adjusted to remain constant and, therefore, addition of density and thickness sensors would be imperative to make the measurement more reliable. In terms of other environmental factors, the moisture measurement is made in a factory environment, where the humidity does not vary greatly. The moisture sensor is intended for applications within the range 40–60% water content. Small ambient humidity changes would be calibrated with the measurement of an empty resonator and remain below noise level i.e., not affect the resonant frequency more than the measurement resolution.

The number of reference samples was increased to 367 during this measurement campaign. Comparing the standard error with $n = 65$ ([16]) and $n = 367$, $(STD/\sqrt{n})$ has reduced from 0.02% to 0.003%. Nevertheless, variation of water layers was high, more than ±1 mm for the same resonant frequencies. Then, at a maximum one standard deviation was allowed from the average height of the equivalent water layer for each dataset and the number of samples was further reduced to 235. Both a linear fit and a parabola fit were determined and they gave $R^2 = 0.78$ and $R^2 = 0.82$, respectively.

4.2. Future Work

There are a number of improvements to the protocol that could be suggested, if this was to be used in the future. In general, the assumption of an equivalent water layer is too simplified, because the thickness and impact of permittivity of the dry material is ignored completely. In reality, the thickness of the material changes randomly as the mat passes the sensor.

Another simplification in the presented measurement configuration is that the resonant frequency shift was measured only once per measurement footprint. The mat was moving constantly through the sensor so it was not possible to take, for example, three recordings at the same location. This would be possible employing only one sensor, if the belt was stopped during the measurement. In general, measurements of reference samples of a very inhomogeneous material in a laboratory do not correlate perfectly with factory tests, even though the performance of the strip line cavity sensor was verified earlier in [12,14,20].

Repeatability of one point-like measurement is poor, as the material changes constantly. Even though the mat has supposedly similar moisture and thickness conditions along the lateral direction, deploying a sensor array of similar sensors that measure and record the resonant frequency shift simultaneously could improve accuracy. Impurities in the material cause a widening of the resonance curve and difficulty determining the exact resonant frequency. Operation of the belt filter press does not allow for a constant thickness of the mat. In conclusion, it would not appear sufficient to assume that the variation in thickness of the biofuel mat has a negligible influence when using a single parameter moisture measurement.

Acknowledgments: The author was with the Department of Radio Science and Engineering, Aalto University, Espoo, Finland, during 2009–2013. Collaboration is acknowledged in particular with Pertti Vainikainen, Tommi Laitinen and Jan Järveläinen.

Conflicts of Interest: The author declares no conflict of interest.

References

1. Singh, D.R. A high speed, high temperature, on-line moisture sensing technique for paper web applications. In Proceedings of the IEEE International Conference on Industrial Technology, Goa, India, 19–22 January 2000; pp. 24–25.
2. Yogi, R.A.; Parolia, R.S.; Karekar, R.N.; Aiyer, R.C. Microwave microstrip ring resonator as a paper moisture sensor: study with different grammage. *Meas. Sci. Technol.* **2002**, *13*, 1558–1562. [CrossRef]
3. Höltta, V. Plant Performance Evaluation in Complex Industrial Applications. Ph.D. Thesis, Helsinki University of Technology, Espoo, Finland, 2009.
4. Fischer, M.; Vainikainen, P.; Nyfors, E. On the permittivity of wood and the on-line measurement of veneer sheets. In Proceedings of the Workshop on Electromagnetic Wave Interaction with Water and Moist Substances, Atlanta, GA, USA, 14–18 June 1993; pp. 347–354.
5. Jakkula, P.; Karhu, J.; Luostarinen, K.; Limingoja, M. Method and Measuring Instrument for Measuring Water Content. U.S. Patent 8,188,751 B2, 29 May 2012.
6. Holmberg, H. Biofuel Drying as a Concept to Improve the Energy Efficiency of an Industrial Chp Plant. Ph.D. Thesis, Helsinki University of Technology, Espoo, Finland, 2007.
7. Jindarat, W.; Sungsoontorn, S.; Rattanadecho, P. Analysis of Energy Consumption in Drying Process of Biomaterials Using a Combined Unsymmetrical Double-Feed Microwave and Vacuum System (CUMV)—Case Study: Tea Leaves. *Dry. Technol. Int. J.* **2013**, *31*, 1138–1147. [CrossRef]
8. Tiuri, M.; Heikkila, S. Microwave Instrument for Accurate Moisture Measurement of Timber. In Proceedings of the 9th European Microwave Conference, Brighton, UK, 17–20 September 1979; pp. 702–705.
9. Nyström, J.; Dahlquist, E. Methods for determination of moisture content in woodchips for power plants—A review. *Fuel* **2004**, *83*, 773–779. [CrossRef]
10. Dinčov, D.D.; Parrott, K.A.; Pericleous, K.A. Heat and mass transfer in two-phase porous materials under intensive microwave heating. *J. Food Eng.* **2004**, *65*, 403–412. [CrossRef]
11. Nyfors, E.; Vainikainen, P. *Industrial Microwave Sensors*; Artech House: Norwood, MA, USA, 1989; pp. 141–145.
12. Nyfors, E.; Vainikainen, P.-V.; Fischer, M.T. Method and Apparatus for the Measurement of the Properties of Sheet- or Foil-Like Materials of Low Electrical Conductivity. U.S. Patent 4,739,249, 23 April 1987.
13. Metriguard Veneer Testers Catalog Brochure Material Internet. Available online: https://www.metriguard.com/category/veneer-testers (accessed on 21 September 2018).
14. Fischer, M.; Vainikainen, P.; Nyfors, E. Design Aspects of Stripline Resonator Sensors for Industrial Applications, Helsinki University of Technology Report S214. *J. Microw. Power Electromagn. Energy* **1995**, *30*, 246–257. [CrossRef]
15. Olkkonen, M.-K. Non-Destructive RF Moisture Measurement of Bio Material Web. Master's Thesis, Aalto University, Espoo, Finland, 2010.
16. Olkkonen, M.-K.; Laitinen, T.; Vainikainen, P. Moisture Measurements of Bio-Material Web Using an RF Resonator Sensor. In Proceedings of the 9th International Conference on Electromagnetic Wave Interaction with Water and Moist Substances (ISEMA 2011), Kansas City, MO, USA, 31 May–3 June 2011; pp. 1–7.
17. Reeb, J.E.; Milota, M.R. Moisture Content by the Oven-Dry Method for Industrial Testing. In Proceedings of the Western Dry Kiln Association Meeting, Portland, OR, USA, 1999; Available online: https://ir.library.oregonstate.edu/xmlui/handle/1957/5190 (accessed on 31 October 2018).
18. Kraszewski, A.W.; You, T.S.; Nelson, S.O. Microwave resonator technique for moisture content determination in single soybean seeds. *IEEE Trans. Instrum. Meas.* **1989**, *38*, 79–84. [CrossRef]

Sensors **2018**, *18*, 3844

19. Meyer, W.; Schilz, W.M. Feasibility Study of Density-Independent Moisture Measurement with Microwaves. *IEEE Trans. Microw. Theory Tech.* **1981**, *29*, 732–739. [CrossRef]
20. Fischer, M.; Vainikainen, P.; Nyfors, E. Dual-mode stripline resonator array for fast error compensated moisture mapping of paper web. In Proceedings of the IEEE MTT-S International Microwave Symposium Digest, Dallas, TX, USA, 8–10 May 1990; pp. 1133–1136.

sensors

MDPI

Article

A Noninvasive TDR Sensor to Measure the Moisture Content of Rigid Porous Materials

Zbigniew Suchorab [1], Marcin Konrad Widomski [1], Grzegorz Łagód [1,*], Danuta Barnat-Hunek [2] and Dariusz Majerek [3]

[1] Faculty of Environmental Engineering, Lublin University of Technology, Nadbystrzycka Str. 40B, 20-618 Lublin, Poland; Z.Suchorab@pollub.pl (Z.S.); M.Widomski@pollub.pl (M.K.W.)
[2] Faculty of Civil Engineering and Architecture, Lublin University of Technology, Nadbystrzycka Str. 40, 20-618 Lublin, Poland; D.Barnat-Hunek@pollub.pl
[3] Faculty of Fundamentals of Technology, Lublin University of Technology, Nadbystrzycka Str. 38, 20-618 Lublin, Poland; D.Majerek@pollub.pl
* Correspondence: G.Lagod@pollub.pl; Tel.: +48-81-538-4322

Received: 30 September 2018; Accepted: 12 November 2018; Published: 14 November 2018

Abstract: The article presents the potential application of the time domain reflectometry (TDR) technique to measure moisture transport in unsaturated porous materials. The research of the capillary uptake phenomenon in a sample of autoclaved aerated concrete (AAC) was conducted using a TDR sensor with the modified construction for non-invasive testing. In the paper the basic principles of the TDR method as a technique applied in metrology, and its potential for measurement of moisture in porous materials, including soils and porous building materials are presented. The second part of the article presents the experiment of capillary rise process in the AAC sample. Application of the custom sensor required its individual calibration, thus a unique model of regression between the readouts of apparent permittivity of the tested material and its moisture was developed. During the experiment moisture content was monitored in the sample exposed to water influence. Monitoring was conducted using the modified TDR sensor. The process was additionally measured using the standard frequency domain (FD) capacitive sensor in order to compare the readouts with traditional techniques of moisture detection. The uncertainty for testing AAC moisture, was expressed as RMSE (0.013 cm^3/cm^3) and expanded uncertainty (0.01–0.02 cm^3/cm^3 depending on moisture) was established along with calibration of the applied sensor. The obtained values are comparable to, or even better than, the features of the traditional invasive sensors utilizing universal calibration models. Both, the TDR and capacitive (FD) sensor enabled monitoring of capillary uptake phenomenon progress. It was noticed that at the end of the experiment the TDR readouts were 4.4% underestimated and the FD readouts were overestimated for 12.6% comparing to the reference gravimetric evaluation.

Keywords: time domain reflectometry; TDR; frequency domain; FD; porous materials; building materials; moisture

1. Introduction

Among the numerous available methods for estimation of porous media water content, the time domain reflectometry (TDR) technique is considered as one of the most useful [1,2]. Contrary to the capacitive methods, including the frequency domain (FD) or resistance-based methods of moisture measurements, TDR allows moisture determination with satisfactory accuracy, regardless of external factors, including e.g., temperature and, to a certain extent, salinity, affecting the obtained results [3–5].

The first historical applications of TDR method in soil water content measurements were reported in the 1980s [6]. Thereafter, the method has been constantly developed and refined, using ongoing achievements in the field of electronics and sensor construction [7–15] as well as the required calibration procedures [6,16,17]. A notable increase in the range of possible applications of TDR method also became apparent [18–20].

The TDR method generally utilizes observations of the electromagnetic pulse propagation time along the sensor placed in the material that moisture is being investigated. The dimensionless apparent permittivity ε, being a measure of molecules' behaviour under the alternating electromagnetic field and energy dissipation of the material after electromagnetic field is released, is a basic, fundamental parameter required for successful TDR application [6,21–33].

Several factors affect the apparent density values of multiphase porous media, including their structure, particle size distribution, etc. but the dipolar character of water molecule makes the influence of water the most important. The electric load distribution for water, resulting in the high value of relative apparent permittivity reaching 80 [-], is different than for the other phases of the porous media [21]. The reported values of dimensionless apparent permittivity for air, granite, sandstone, clay and sand were equal to 1, 4−9, 2−3, 2−6, 4−5, respectively [21].

The dielectric permittivity of the materials is a complex number, consisting of a real (ε') and an imaginary (ε'') part. The real part describes the base value for moisture estimation using the TDR technique, i.e., the amount of released energy in the alternating field, while the imaginary part covers energy loses due to the ionic conductivity, highly dependent to salinity of the medium [23]. The complex dielectric permittivity of saline medium may be calculated according to the following formula [4,22]:

$$\varepsilon_\omega = \varepsilon'_\omega - i\left(\varepsilon''_\omega + \frac{\sigma_0}{\varepsilon_0 \omega}\right) \tag{1}$$

where: ε'_ω—real part of dielectric permittivity of medium at ω frequency [-], ε''_ω—imaginary part of dielectric permittivity of medium at ω frequency [-], i—imaginary unit ($i^2 = -1$), σ_0—electrical conductivity [S/m], ε_0—dielectric permittivity of vacuum ($\varepsilon_0 = 8.85 \times 10^{-12}$ F/m), ω—angular frequency of the external electric field [1/s].

The above formula explains that the imaginary part influences measurements in low frequencies of electromagnetic field, e.g., applied in the FD method. The operating frequency of many of the TDR multimeters reaches values of approx. 1 GHz [4], high enough to minimize the influence of imaginary part on the value of complex dielectric permittivity of a saline medium. Thus, it can be assumed that the ionic conductivity has low effect on the TDR readouts, which may be stated as one of the most important advantages of this method in relation to the others based on resistance and capacitance. It must be underlined here, that salinity influences the responses of the reflectometric traces—amplitude of the pulse diminishes and the measuring peaks are flatten, which in some cases may result in the decrease in information available from the tested medium affecting the measuring accuracy [4]. On the other hand, it should be also mentioned, that amplitude diminishing, recognized as a negative phenomenon in the TDR measurement, can be also utilized to evaluate medium salinity of the medium, basing on suitable, insightful waveform interpretation [24].

Therefore, the following formula may be used to calculate the relative apparent permittivity of the porous material [9]:

$$\varepsilon = \left(\frac{c \cdot t_p}{2L}\right)^2 \tag{2}$$

where: c—light velocity in vacuum [m/s], t_p—travel time along the TDR sensor [s], L—length of measuring elements of the TDR sensors [m].

Moisture measurement using the TDR method rely on the determination of the electromagnetic pulse travel time along the rods of the TDR probes (Figure 1), which generally consist of a concentric cable, head and measuring rods buried into the tested material. The readouts are based on the reflections on particular discontinuities of the sensor waveguide, being the elements of its construction. Usually, the described discontinuities are located at the beginning and the end of probe. Figure 1 shows an exemplary TDR probe, with black arrows marking the discontinuities of waveguide for the electromagnetic pulse. During measurement the rods have to be inserted into the tested material. The contact between the rods and tested material should be precise and permanent to allow the reliable readouts.

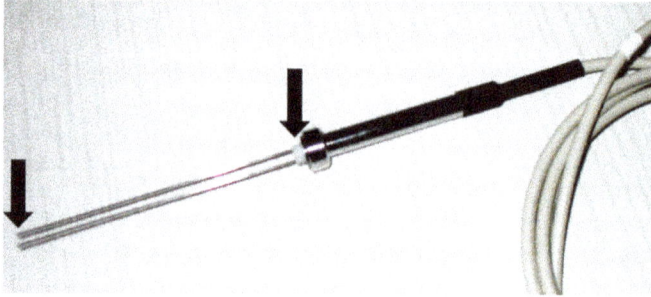

Figure 1. LP/ms probe (ETest, Lublin, Poland) for moisture determination using TDR method.

The TDR technology utilizes two types of pulses emitted by the pulse generators: the step pulse and the needle pulse. Both differ in the length of the incident pulse, in the first case the emitted pulse is wider in comparison to the needle pulse length. The exemplary waveforms obtained by the TDR multimeter utilizing 300 ps rise-time needle pulse generator [11] are presented in Figure 2.

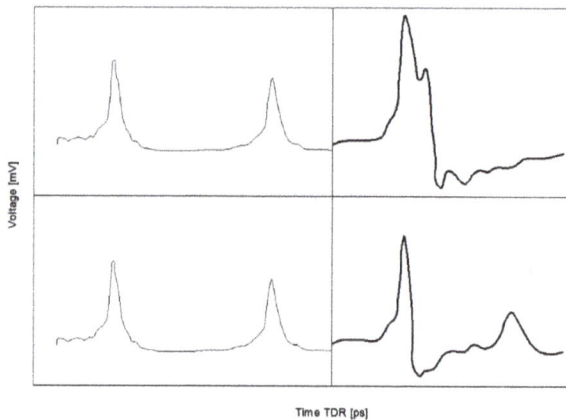

Figure 2. Example of TDR waveforms for dry (**top**) and moist (**bottom**) material acquired from an ETest LP/ms TDR probe (own elaboration based on calibration tests). Left-hand side—control peaks, right-hand side—measuring peaks.

They represent the responses of TDR probe on dry and wet material tested by the TDR LP/ms probe (ETest, Lublin, Poland) where the upper trace is representative for the dry materials with low value of the apparent permittivity and the bottom trace represents wet material. The visible and marked differences between peaks of the TDR traces for dry and wet material depend on the apparent permittivity values of the tested material. The left-hand side of both waveforms presents testing peaks, which are not influenced by material moisture. The right-hand side peaks are the measuring ones. The first, positive measuring peak is constant for both, dry and moist, materials. It represents the reflection from the probe input (right black arrow in Figure 1) and the second one, with smaller voltage, representing the reflection from the end of the probe (arrow at the end of the rod). Distance between both measuring peaks expressed in time can be recalculated into the apparent permittivity using Equation (2). The longer time of signal propagation for wet material results in shifting the second measuring peak towards the right side of the graph. Determination of moisture in porous materials based on the measured dielectric permittivity can be accomplished using various theoretical and physical models [25–27] or the empirical calibration formulas obtained by the experimental examinations [6,16,17,28].

The significant advantage of the physical models is their independence from calibration procedures. On the other hand, the most essential of their disadvantages is the complicated mathematical description hindering laboratory measurements. The physical descriptions of dielectric parameters of porous materials as ternary mixtures were elaborated in 1892 by Rayleigh [34], in 1904 by Maxwell Garnett [35], and in 1946 by Polder and van Santen [36]. Among the present dielectric models of porous media there should be mentioned the models by De Loor [25], Tinga [26], Roth [37], Whalley [38] and Noborio [28]. All the above mentioned models differ in the approach to the porous material structure, geometry, shape and morphology of the grains but also differ in the grade of complexity and, as it was mentioned above, they are not easy for the practical aspects of moisture evaluation in the real laboratory or in-situ conditions.

The other approach of calibration of the TDR probes for moisture determination is to describe the dielectric parameters of the moist porous media and to develop an empirical model based on laboratory tests allowing the correlation between the gravimetric and TDR moisture readouts. Among the empirical models universal and individual models can be distinguished. The universal models are developed on the base of the multiple investigations of numerous media to describe various materials that differ in density, porosity and structure solid phase. The individual models are elaborated to find calibration formula for the particular material or even sensor.

Among most cited universal empirical models two the most important should be mentioned: Topp's [6] and Malicki's [16] formulas. The first is the third order polynomial function relating the moisture of porous material to only one measured parameter – apparent permittivity. This enables quick estimation for many porous materials without the prior calibration independently on the examined material and sensor used. On the other hand, this method not always provides the correct results of the readouts. According to Schapp et al. [39] the possible uncertainty of measurement can vary in the range between 0.05 and 0.15 cm^3/cm^3, which may be caused by the differences of solid phase structure of the examined material. According to Černý [11], standard uncertainty of moisture estimation by the Topp's model equals 0.0468 cm^3/cm^3. Additionally, it should be considered that Equation (3) is applicable for porous media with bulk density close to 1500 kg/m^3, only for volumetric moisture content below 0.5 cm^3/cm^3 and should not be used for organic soils or mineral soils containing organic material and clay [24,40].

Malicki's approach improves the accuracy of moisture determination using the TDR technique compared to the Topp's model and extents its application. It is described by the semi-empirical formula considering bulk density of the tested material, a part of the apparent density.

The semi-empirical models are still universal and present the acceptable accuracy making them common in reflectometric investigations. On the other hand, many individual calibration formulas elaborated for the particular materials or sensors and offering the better accuracy than empirical models by Topp and Malicki may be found in the literature [19,41–46].

2. Concept of the Surface TDR Sensor

The building materials present the large share among the porous media. In moderate climate the housing sector suffers from deteriorating water presence inside the building envelopes. Water present in porous building materials decreases their bearing as well as the thermal properties, negatively influencing energetic performance of the buildings. Another negative effect of water presence in the building envelops is the risk of microbial threat and Sick Building Syndrome (SBS) symptoms.

Construction of the previously described traditional TDR probes significantly constricts moisture measurements in the firm porous media, including most of the building materials. The above is triggered by the geometrical and mechanical properties of measurement units—steel rods. They are usually quite long and thin, and, like in case of the LP/ms probes made by ETest, also frail. Such probes are useful during measurements of soil moisture, but in case of water content determination for hard building materials they are inapplicable.

Thus, most of the reported studies concerning water content of building materials were performed under the laboratory conditions allowing the proper preparation of samples. The preparatory activities usually covered drilling the pilot holes in which the rods of the probe were inserted or drilled holes of larger diameter in which void air space was filled with the drilling dust [47–50]. Unfortunately, all these procedures were altering the structure of studied material, including its water characteristics. Thus, the obtained readouts for the transformed material were not reflecting the real moisture conditions of the studied sample. There are two possible concepts of solving the problem of moisture measurements in firm building materials:

- construction of TDR probes of significant size, consisting of steel rods of the required diameter and durable head [4];
- construction of the TDR surface sensor.

The first surface TDR sensor concepts were reported in the 1990s [14,15]. The probe proposed by Selker et al. [14] utilized the long brass wire, shaped in the spiral manner, covered by acrylic plate. The individual ε-θ calibration was required for this probe and its measurements uncertainty wearied in the range ± 0.02 cm^3/cm^3. On the other hand, the idea of sensors proposed by Perrson and Berndtsson [15] was based on application of the typical three-rod probes covered by the properly carved dielectric of known thickness and dielectric characteristics, allowing determination of dielectric parameters, thus water content, of medium located below the cover. This solution was rather primitive but it enabled, to some extent, the non-invasive determination of moisture in porous medium.

The interesting and different solution of surface probe was proposed by Wraith et al. [51] as the probe for determination of moisture in top soil. The probe similar to sledges could be pulled over the soil surface like the georadar, allowing measurements of top soil water content.

Ito et al. [52] proposed the multi-TDR probe, allowing measurements of evaporation from soil surface, consisting of the layered composite of glass and resign, covering 17 copper electrodes in shape of stripes, 100 mm length, 0.02 mm width and 0.01 mm thick. The unit was consisting of 8 combined probes, for which the individual calibration was required.

The new concept of non-invasive TDR sensors was proposed by Choi et al. [53]. The three-rod surface probe was additionally equipped in the piezoelectric sensor and accelerometer, allowing the measurements of dry bulk density, soil moisture and modulus of elasticity. All the measurements may be performed without altering the soil surface.

The concept of the TDR surface sensor for firm materials was presented in the patent reservation [54]. The prototype and possible applications were already reported [4]. Modifications of this TDR sensor allowing moisture measurements in firm porous materials of irregular surface were also presented in patent's documentation [55–57].

The performed literature studies showed that time domain reflectometry is a very applicable technique in the field of soils science. However, it was also indicated that there is a need for development of the method allowing TDR application in determination of wall barriers moisture conditions. Thus, construction of the probes and development of the required measurement methodology are required.

The aim of studies presented in this paper was to apply the indirect moisture detection technique, TDR with the modified sensor construction, to determinate the unsaturated water flow in a rigid, porous building material. The conducted research was additionally supplemented with the FD (capacitive, non-invasive probe and direct gravimetric evaluation). The aerated autoclaved concrete (AAC) was selected as a tested material, due to popularity of the material in modern building sector but also the proper hygric parameters that would reveal the measuring potential of the tested sensors: low density, high porosity, high capillarity, saturation, etc.

3. Materials and Methods

3.1. Details of the Developed Sensor

The subject of study was a TDR surface sensor developed to examine moisture content of rigid porous media as building materials, building barriers or rocks. The prototype specimen applied for the presented research was constructed of black polyoxymethylene (POM)—plastic characterized by good mechanical parameters including strength, stiffness, ductility and value of apparent permeability at the level of 3.8 [7]. The length of the probe and waveguide was equal 200 mm, while width 50 mm. Measuring elements were manufactured from brass flat bar 2 mm × 10 mm. The device was equipped with a cylinder shaped handle. Communication between the probe and the TDR multimeter was provided by the BNC connector with simple printed circuit consisting of the two lines soldered to the pins of the BNC connector and both bars on the other side. There was a resistor soldered between two lines. A schematic view of the proposed surface TDR sensor construction is presented in Figure 3.

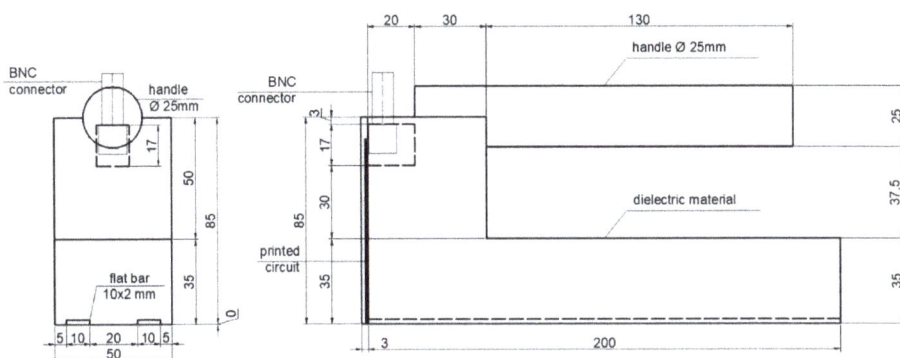

Figure 3. Schematic view of the proposed sensor construction.

Figure 4 presents the TDR traces obtained with the discussed sensor.

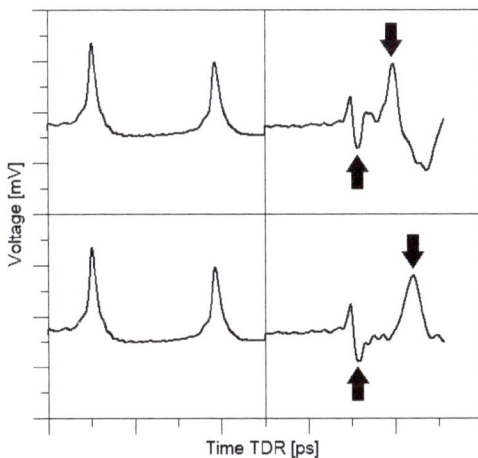

Figure 4. Electric response of the developed sensor for different environments: upper trace—air-dry sample, bottom trace—saturated sample.

The first negative peak (marked with an arrow) is constant in position and is a consequence of mounting of the resistor in the printed circuit of the sensor. The second, positive peak (marked with arrow) means the measuring element termination and its position results from material moisture.

Before the sensor was used for laboratory experiments it was tested to define the range of electromagnetic signal influence in the measured material. From the literature it is known that this mainly depends on the spacing between the two measuring waveguides [58]. This parameter of the developed sensor was evaluated in laboratory conditions, methodology of its estimation was described in the following article [59]. In case of the described sensor the range of signal influence was defined as 40 mm deep.

3.2. Measuring Setup

The following materials were applied to the experiment:

- Aerated concrete, dry apparent density 600 kg/m^3;
- Laboratory oven VO-500 (Memmert, Schwabach, Germany);
- Bitumen isolation;
- Laboratory scale WPT 6C/1 (RADWAG, Radom, Poland);
- Multifunctional scale WPW 30/H3/K (RADWAG, Radom, Poland),
- Water reservoir equipped with necessary equipment to sustain the constant water level;
- TDR equipment including laboratory multimeter LOM (ETest, Lublin, Poland);
- TDR sensor presented in this article, concentric cable;
- Personal Computer for meter control and data management;
- Capacitive moisture meter LB-796, (LABEL, Reguły, Poland);
- Atomizer (for calibration procedure).

3.3. Preliminary Research

Preliminary research was conducted to establish the basic physical and hygric parameters of the materials, important from the point of view of the conducted experiment: apparent density of the material and its saturated water content. Three samples 50 mm × 50 mm × 45 mm were prepared and dried in the 105 °C in the laboratory oven. After dry mass was determined, the samples were saturated to allow determination of gravimetric and volumetric water contents. Gravimetric and volumetric water content were determined using the following equation [29]:

$$w = \frac{m_n - m_s}{m_s} \tag{3}$$

$$\theta_V = \frac{V_w}{V_{tot}} \tag{4}$$

where: w—gravimetric water content [kg/kg], m_n—mass of wet sample [kg], m_s—mass of dry sample [kg], θ_V—volumetric water content [cm^3/cm^3], V_w—volume of water [cm^3], V_{tot}—total volume of the sample [cm^3].

3.4. Calibration the Sensor

Unusual sensor construction and insufficient verification of calibration formulas, intentionally developed for soils, caused the necessity of the individual calibration procedure for the newly developed sensor which was going to be applied for building materials.

The dimensions of the samples was the following: 220 mm × 120 mm × 40 mm. External surfaces of the samples were polished to provide equal adherence to the tested material. The first step was conducted on the dry set of the samples. Then the samples were sequentially moistened using atomizer with steady portions of water to achieve the full saturation. During the experiment the samples were weighed using laboratory scale and volumetric water content was evaluated using the Equation (4). Then the surface sensor was pressed to the tested sample with constant pressure and the effective dielectric permittivity was read. For the statistical post-processing, each step of measurement was repeated five times.

3.5. Model of Regression

According to the authors' experience and the literature [41,42] the assumed general form of calibration equations (see Equation (5)) has a second order polynomial function character. The input data for the model covered volumetric water content obtained due to the direct gravimetric measurements and the mean value of the effective dielectric permittivity obtained by the reflectometric measurements:

$$\theta = \beta_0 + \beta_1 \cdot \overline{\varepsilon_{eff}} + \beta_2 \cdot \overline{\varepsilon_{eff}}^2 + \epsilon \tag{5}$$

$$(p) \quad (p) \quad (p)$$

where: θ—volumetric water content determined by polynominal model [cm^3/cm^3]; $\overline{\varepsilon_{eff}}$—mean effective dielectric permittivity obtained by reflectometric measurements [-], ϵ—random error of normal distribution, p—critical level of significance (* $p < 0.05$; ** $p < 0.01$; *** $p < 0.001$).

3.6. Calculation of Uncertainty

The measurements uncertainties type A determine the quality of models' fitting to the experimental data. The source of type B of uncertainties are the measuring uncertainties of the instruments used within the calibration procedure. In the Equation (5) there are two sources of variance. The first is the regression uncertainty and $\varepsilon \sim N(0, \sigma)$ which comes from the randomization which results in some variability of all estimated regression parameters, that is expressed in covariance matrix $\sigma^2(X'X)^{-1}$. The second source is a consequence of fact, that it is not possible to reveal the true dependence between the selected predictors and the dependent variable—fitted volumetric water content value may differ from the true value of θ because it is impossible to control all variables affecting it [60]. Uncertainties type B are neglected from the investigation because they are of lower level comparing to the uncertainty of A type. In the assumed model four factors affect measurement uncertainty: estimators β_0, β_1, β_2, and the dielectric permittivity:

$$\theta = f(\beta_0, \beta_1, \beta_2, \varepsilon) \tag{6}$$

Using the error propagation law, the combined standard measurement uncertainty (including uncertainties type A and B) may be presented as follows [60,61]:

$$u_C(\theta) = \sqrt{\left(\frac{\partial \theta}{\partial \varepsilon} u(\varepsilon)\right)^2 + \sum_{i=0}^{2}\left(\frac{\partial \theta}{\partial \beta_i} u(\beta_i)\right)^2 + 2\sum_{i=0}^{2}\sum_{j=i+1}^{2}\frac{\partial \theta}{\partial \beta_i}\frac{\partial \theta}{\partial \beta_j} u(\beta_i, \beta_j)} \tag{7}$$

so:

$$u_c^2(\theta) = S^2\left(1 + \frac{1}{n} + \sum_i \left(\frac{\partial \theta}{\partial \beta_i}\right)^2 u^2(\beta_i) + 2\sum_{ij}\left(\frac{\partial \theta}{\partial \beta_i}\frac{\partial \theta}{\partial \beta_j}\right)cov(\beta_i\beta_j)\right) \tag{8}$$

The expanded measurement uncertainty was determined using the following formula:

$$U(\theta) = k_p \cdot u_c(\theta) \tag{9}$$

where: k_p—coverage factor, calculated from t-student distribution for $\alpha = 0.05$, depending on the number of degrees of freedom, it oscillates around 2.

3.7. Capillary Suction Test

The aim of the laboratory research was to assess the measurement potential of the prototype TDR sensor and to demonstrate its applicability in practical aspects as well as to compare its measuring features with a popular moisture sensor available on the market. The research was focused on monitoring of water transport in the model aerated concrete wall barrier utilizing newly developed TDR sensor and the Label LB-796 capacitive FD sensor (LABEL, Reguły, Poland), providing satisfactory readouts of moisture in building barriers, being successfully used for expertises concerning water damage of the buildings. The applied sample was cut from the concrete block to the dimensions of 240 mm × 240 mm × 350 mm, dried to constant mass and covered with thin layer of the bitumen isolation in order to minimize the environmental impacts on the studied process. The scheme of capillary rise monitoring in aerated concrete is presented in Figure 5. The studied sample was inserted approx. 1 cm below the distilled water surface in the reservoir. The water level was kept constant with the help of the glass tube filled with water. The measurements points were assigned in 5 cm interval (levels 5, 10, 15, 20, 25 and 30 cm) above the water surface. During the measurement the TDR sensor was carefully contacted to the tested sample maintaining the constant pressure. The middle of its width was positioned at the particular measuring levels as visible in Figure 5. The similar procedure was repeated for the applied FD sensor, according to the producer's recommendations.

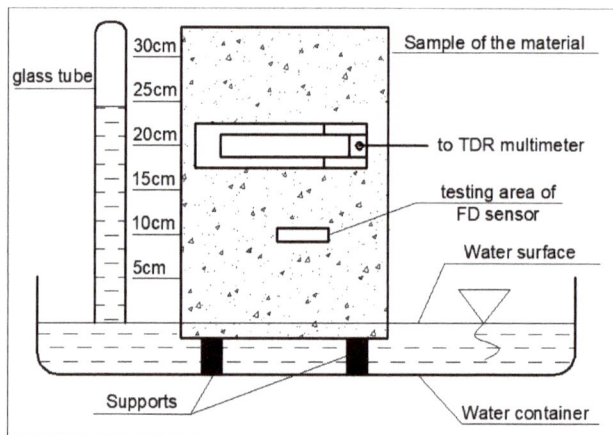

Figure 5. Schematic view of the capillary uptake setup.

Before and after the capillary suction test the sample was weighed, which enabled verification of the indirect readouts (by the tested surface TDR and capacitive sensors) to the direct readouts obtained gravimetrically. The mass of sample dried in temperature of 105 °C was equal to 12.39 kg while the determined apparent density reached the value of 614.6 kg/m³ and was higher than declared by the producer. The determined mass of the moist concrete sample after capillary rise test was equal to 16.91 kg, thus the increase of 4.52 kg was observed in relation to mass before the test.

The TDR and LB-769 studies were performed for a time duration of 16 days. Moisture readouts were performed three times a day. Each measurement was based on contact of the TDR and capacitive sensors with the studied sample at the given height. During the measurement a constant pressure between both sensors and the tested material was kept to minimize the influence of uneven contact condition on measuring accuracy. Each measurement was repeated three times, in order to assure the statistically required number of data. The environmental conditions of measurements were as follows: temperature 20 °C ± 2 °C, relative air moisture 50% ± 5%. The curves presenting the dynamics of capillary rise process by the sample of tested aerated concrete were obtained as the result of the experiment. The values of effective permittivity ε_{eff} were converted into the volumetric moisture content, using the calibration formula obtained within the calibration test (Sections 3.4 and 3.5). The FD sensor readouts were performed in the similar manner, simultaneously to the TDR measurements, according to the producer's guidelines. The FD meter was pre-calibrated by the producer, which enabled reading ready values of moisture content. Since the FD results are presented as the gravimetric water content, the readouts were converted to volumetric water content.

4. Results

4.1. Preliminary Test Results

The basic hygric parameters of the materials were established with the preliminary tests. Apparent density, volumetric and gravimetric saturated water content of the studied material are presented in the Table 1.

Table 1. Basic physical properties of the examined material.

Apparent Density [kg/m³]	Saturated Volumetric Water Content [cm³/cm³]	Saturated Gravimetric Water Content [kg/kg]
612.2 ± 11.2	0.363 ± 0.007	0.593 ± 0.007

4.2. Calibration of the TDR Sensor

With the calibration procedure the dependence between effective dielectric permittivity (read by the TDR surface sensor) and the volumetric water content was achieved. The results are presented in Figure 6a.

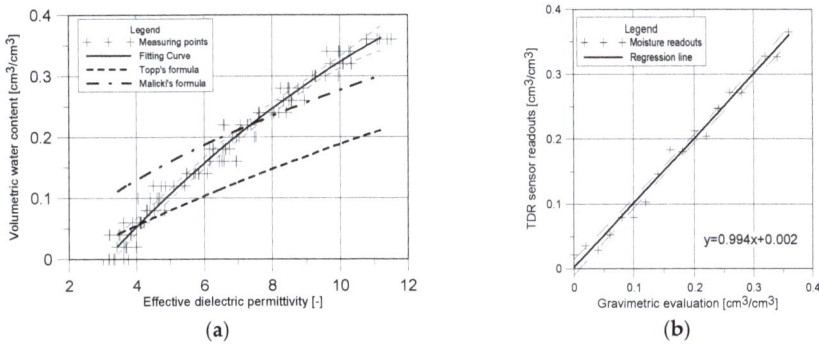

Figure 6. Calibration test results: (**a**) dependence between effective dielectric permittivity and material moisture, (**b**) comparison of data obtained gravimetrically and by reflectometric evaluation.

As it was previously mentioned, the dependence between data presented in Figure 6a can be described using the second order polynomial regression model proposed as Equation (10):

$$\hat{\theta} = -0.1956 + 0.0691\varepsilon_{app} - 0.0017\,\varepsilon_{app}^2 \tag{10}$$

$$(***) \qquad (***) \qquad (*)$$

The basic statistical parameters of the developed regression formula are stated in Table 2.

Table 2. Statistical parameters of the developed calibration model of the surface TDR probe.

Determination Coefficient R^2	Residual Standard Error RSE [cm^3/cm^3]	Root Mean Square Error RMSE [cm^3/cm^3]	F-Model Linearity Test Statistic
0.986	0.014 (df = 16)	0.013	580.752 *** (df = 2; 18)

*** $p < 0.001$.

The comparison of results obtained by model for the surface TDR probe and the gravimetric measurements were presented as graph in Figure 6b.

4.3. Combined Standard and Expanded Measurement Uncertainty

The results of determined (according to Equations (8) and (9)) combined standard and expanded measurements uncertainties were presented in Figure 7.

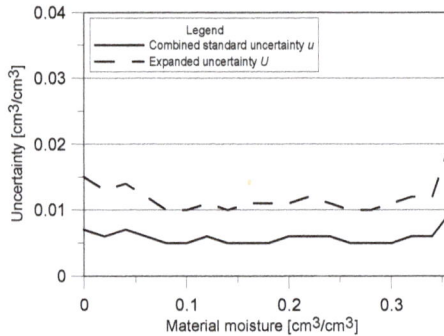

Figure 7. Combined standard and expanded measurements uncertainties of the TDR surface sensor for aerated concrete.

For most of the material moisture range the expanded uncertainty of TDR measurement using the surface sensor is about 0.01 cm^3/cm^3. Only in nearly dry and saturated conditions its value is higher, 0.015 and 0.02 cm^3/cm^3, respectively.

4.4. Capilary Suction Results

The graph presented in Figure 8 shows the curves of capillary rise determined by the applied surface TDR sensor in the reference points at given heights above the water level. It represents the mean values of three repetitions, supported by the standard deviations expressed as error bars.

The mean values obtained by the applied FD sensor and supported by SDs values are presented in Figure 9.

The changes of moisture in subsequent reference points determined by the indirect electric measurements were observed. The initial water content presented in Figures 8 and 9 showed values close to zero, and was equal 0.01 cm^3/cm^3 and 0.02 cm^3/cm^3 when determined by the TDR and FD sensors, respectively. The reported initial readouts slightly greater than zero may be caused by the manner of sample preparation (drying in 105 °C to constant mass) and measurement uncertainty of applied sensors for the assumed model of regression.

The readouts of both applied sensors showed a very fast increase in moisture at the 5 cm reference point. The trend of the increase is clear, which underlines the strong capillary properties of the tested medium. At the beginning of the second day of the experiment the full saturation of medium by water was observed in point located at the height of 5 cm. The increase in water content at higher level was also rather rapid but was shifted in time—at the height of 10 cm the presence of water was noted on the another day of experiment. The discussed increase was also dynamic and within the next day the full saturation conditions were achieved. Increase in water content at the height of 15 cm was less dynamic, comparing to the lower heights. The first presence of capillary water was noted after three days of the experiment, while the full saturation was observed four days later. The fourth reference point, at the height of 20 cm above the water level, showed increase in water content after six days of experiment. Then, the slow gain of moisture was observed, leading to conditions close to full saturation after the next four days. At the height of 25 cm, both sensors (surface TDR and FD), showed increase in water content after 300 h, i.e., after over 12 days, but the differences in reported values, reaching 0.1 cm^3/cm^3 for the TDR sensor and 0.2 cm^3/cm^3 for FD one, are visible. No increase in water content was observed by both of the probes in the reference point at the height of 30 cm.

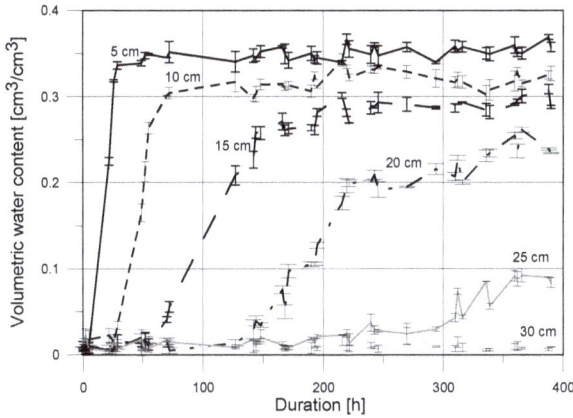

Figure 8. Capillary rise determined by TDR surface sensor in sample of aerated concrete.

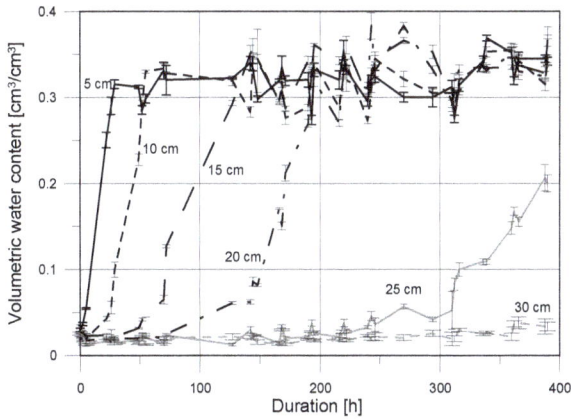

Figure 9. Capillary rise determined by FD capacitive sensor in sample of aerated concrete.

5. Discussion

5.1. Discussion on the Calibration Results and Uncertainty Calculations

According to the assumed model of regression, the second order polynomial formulas were obtained. In order to underline the differences between values of dielectric permittivity obtained by the measurements using the developed TDR sensor and the application of typical invasive probes, there were presented values of water content for the respective values of the dielectric permittivity obtained by formulas by Topp [6] and Malicki [16].

In case of dry samples and moisture contents below 0.05 cm^3/cm^3 the effective apparent permittivity determined using the surface TDR sensor reaches the values in the range between 3 and 4. This is the consequence of the values of apparent permittivity of solid phase of material and apparent permittivity of polyoxymethylene that equals 3.8. In the higher ranges of moisture the readouts of apparent permittivity by the TDR surface sensor show the greater moisture than values read by the traditional invasive probe using Topp's or Malicki's calibration formulas. This is mainly caused by the influence of the polyoxymethylene covering the waveguides and significantly decreases the effective apparent permittivity read by the surface sensor at the particular level of the sample. The estimated calibration Equation (12) considers this influence and precisely reproduces the dependence between the examined moisture and readouts of apparent permittivity by the surface TDR sensor. This is also confirmed by the statistical characteristics of the applied model, mainly coefficient of determination which equals 0.986 and Residual Standard Error (RSE) = 0.014 cm^3/cm^3. Also, the linear formula of regression presented in Figure 6b has the following features: slope value equal 0.994 and y-intercept value equal 0.002. Levels of significance of particular parameter estimators in the Equation (10) are lower than 0.001 in case of β_0 and β_1. Only in the case of estimator β_2 the significance level is below 0.05. Simultaneously, the analysis of F Statistic ($p < 0.001$) confirms the statistical significance of the applied model. Root mean square error (RMSE), the frequently used measure of uncertainty, equals 0.013 cm^3/cm^3 and is lower that could be found in the literature concerning even the invasive probes. According to the data presented by Ju et al. [46] using the Topp's model in relation to the selected soils caused uncertainties expressed as RMSE in the range of 0.01–0.066 cm^3/cm^3. The RMSE value for the model proposed by Roth et al. [37] was in the range of 0.008–0.037 cm^3/cm^3 depending on material, while the RMSE for moisture estimation using the Malicki's model [16] equals 0.03 cm^3/cm^3. The RMSE value obtained for the described surface TDR sensor is smaller than presented in the cited literature, anyway it must be remembered that discussed formulas are universal and because of that the quality of data fitting may be worse. The model presented in this article is individual, dedicated to the particular sensor and material, which may explain the better projection of the discussed dependence ε_{eff}-θ. Analyzing the RMSE value established for the presented sensor it should be mentioned that it is located in the range of RMSE values established by Udawatta et al. [42] for individual models estimated for traditional invasive probes and different materials (0.008–0.034 cm^3/cm^3).

Concerning uncertainty determination it must be mentioned, that like RMSE, the obtained values of uncertainties are lower comparing to the traditional invasive sensors calibrated with the standard empirical formulas. As it was mentioned in Section 4.3, lower and higher ranges of moisture are characterized with the greatest value of measuring uncertainty and the lowest values are noted in the middle range of moisture values available for the tested material. This is the feature of most measuring devices, in this particular case it is caused by the applied model of regression [62]. According to Topp et al. [63] and Amato and Ritchie [64] the uncertainty of measurement ranges between 0.022 and 0.023 cm^3/cm^3, according to Černý [11]—0.0269 cm^3/cm^3, Malicki et al. [16] 0.004–0.018 cm^3/cm^3 and finally Roth et al. [37] 0.011–0.013 cm^3/cm^3. The expanded uncertainty obtained for the presented noninvasive sensor and the tested material is within the values declared by the cited authors for the traditional invasive sensors or even lower. In the opinion of the authors of this elaboration, the beneficial measuring parameters of the prototype non-invasive TDR sensor are the consequence of the following reasons:

- model of regression is individual;
- most of the cited models were developed for soil media, less homogenous in comparison to the tested building material (autoclaved aerated concrete).

5.2. Discussion on Capillary Uptake Experiment Results

Progress of capillary uptake examined using the two applied indirect techniques was presented in Figures 8 and 9 and commented in Section 4.4. Figure 10 shows the comparison of moisture readouts in the individual reference points obtained by the surface TDR and FD sensors. It is visible, that the presented curves are similar for the measurements performed close to the water table level.

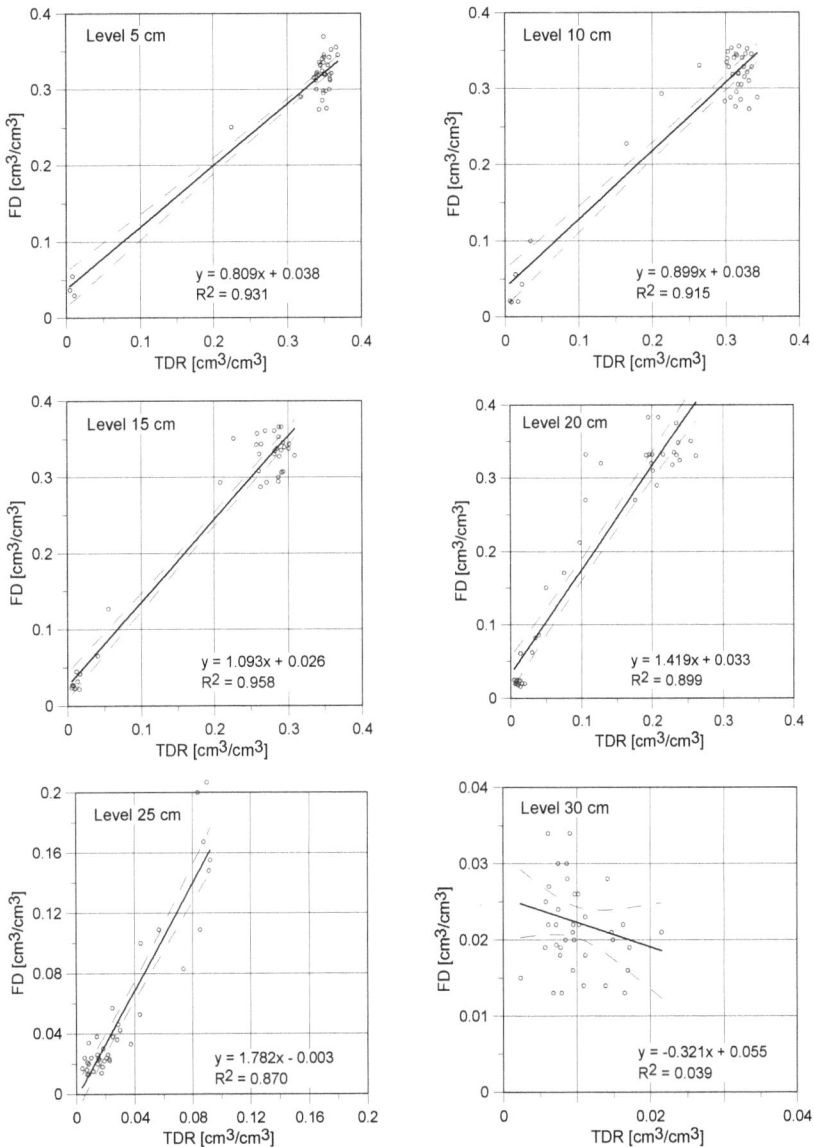

Figure 10. Comparison of capillary rise measurements results obtained by surface TDR sensor and FD capacitive sensor for aerated autoclaved concrete.

For the reference levels at 5, 10 and 15 cm the obtained slopes of linear regression equation were equal to 0.809, 0.899 and 1.093, respectively. Thus, for the first two points the applied TDR noninvasive sensor reports higher values of moisture. Contrary, in case of the third point (15 cm), higher values were shown by the FD capacitive sensor. It should be also noticed that y-intercepts values are slightly above zero in most circumstances, which means that the FD sensor shows higher moisture values under dry conditions. Since this level, the differences between both readouts are more visible. At the height of 20 cm the obtained directional coefficient was equal to 1.419. So, the FD sensor reported water content significantly higher than the TDR. This tendency was noted also for the reference points located in higher elevations, the directional coefficient for linear regression was equal to 1.782 at the level of 25 cm. The huge differences between the TDR and FD sensors were also observed for the reference point at the level of 30 cm (the negative value of regression coefficient).

These differences are probably related to low values of the compared readouts (both sensors reported values from range 0–0.04 cm^3/cm^3) and high measurement uncertainty for the low range of determined water content. All observed differences between both techniques of moisture detection are the consequence of their indirect character and the potential influence of some disturbances which not always could be minimized or eliminated, for example ionic conductivity, contact condition and nonhomogeneous degree of material saturation.

The laboratory studies performed with the application of two types of sensor showed the close moisture readouts and the similar trends of water content changes, both, in its dynamics and quantitative aspect. The following differences were observed:

- moisture readouts at points located at low height about the water table (5 and 10 cm) were higher for the TDR noninvasive sensor;
- at the height of 15 cm moisture content determined by capacitive probe was slightly higher that one indicated by the TDR sensor which is confirmed by the slope of regression higher than 1 and positive value of the y-intercept;
- for low saturation conditions the FD probe showed higher moisture readouts than the TDR surface sensor;
- both of the tested probes showed high measurement instability for low saturation (close to dry), which is visible in Figure 10 for the reference level at 30 cm, with the negative coefficient of regression;
- the maximal noted standard deviation for the TDR sensor was equal to 0.012 cm^3/cm^3 with the maximal standard deviation for the FD probe was higher, reaching 0.037 cm^3/cm^3.

The comparison of sample's mass before and after experiment was performed by the standard procedure and showed the difference of 4.52 kg. In case of the applied electric methods the amount of absorbed water was determined by integration of moisture profile observed at the end of the experiment. The following formula was applied:

$$m = 0.001 \cdot a \cdot b \int_0^h \theta(h) dh \tag{11}$$

where: m—mass of water absorber by the tested material [kg]; a, b—dimensions of sample: width and depth (24 cm); h—height of the sample; $\theta(h)$—water profile for the final time duration of the experiment.

It was determined that for the final part of the experiment the increase in water mass determined by the TDR surface sensor was equal to 4.32 kg and 5.09 kg for the capacitive probe. Thus, the increase in water mass estimated with application of the TDR and FD probes was 4.4% lower and 12.6% greater, respectively, than the increase obtained by the gravimetric method. Calculated underestimation of the increase in water mass by the TDR surface sensor may be related to its range of signal influence, equal to 4 cm, while the thickness of the sample was equal to 24 cm. Assuming the heterogeneous structure of tested material and complex process of water transport, it may be accepted that some part of water was unavailable for the TDR and FD sensors impulse. On the other hand, moisture overestimation presented by the FD probe may be influenced by salt ions present in water inside the tested porous material.

Due to the unique prototypes of probes, different physical characteristics of tested material and its heterogeneity, it is hard to relate the obtained results to the literature reports. The TDR technique is being actually introduced to measurements of water content in rigid porous building materials, so a few literature reports allowing comparison of the results are available. Moreover, the reported results concerning moisture changes in samples of building materials were obtained by the invasive or direct methods.

The measurements of capillary rise in the sample of aerated concrete utilizing the invasive TDR probes were performed by Hansen [47] and during the earlier studies by Suchorab et al. [48]. The aerated concrete researched by Hansen [47] had apparent density of 500 kg/m^3, lower than tested in this paper, so installation of the invasive TDR probes could be easier. The probes were installed at the following heights over the water table 15, 30, 45, 60, 75 and 90 mm, lower and with smaller spacing than applied in our research. Thus, the first registered readouts of water content for the lowest level, 5 cm above water level, were observed after approx. an hour and after 5 h the conditions near full saturation were noted. The increase in water content for higher levels were observed respectively later. To compare the dynamics of the studied processes, the readouts of water content at the height of 90 mm reported by Hansen [47] and 100 mm obtained during the presented studies were analyzed. In case of the surface TDR probe the appearance of water was observed after approx. 20 h and the full saturation after approx. 80 h, while the comparable values were reported by Hansen [47] after approx. 60 and 100 h, respectively. But, the full saturation was probably not achieved, because lower sensors showed higher values of moisture readouts in several points.

The another quoted paper [48] presented results of the similar studies concerning monitoring of capillary rise in aerated concrete by the invasive field ETest FP/mts TDR probes. In this study, the density of applied concrete sample (24 cm × 16 cm × 6 cm) was equal to 500 kg/m^3. The initial conditions showed volumetric water content at the level of 0.1 m^3/m^3. The TDR FP/mts probes were installed at the heights of 5, 10, 15 and 20 cm above the water table, similarly to the experiment concerning application of the FD and surface TDR sensors. The reported experiment lasted 20 days. The maximum value of water content at the end of the experiment was equal to 0.34 cm^3/cm^3 and was comparable to readouts by the surface TDR and FD sensors (0.357 and 0.338 cm^3/cm^3, respectively). The increase in moisture to full saturation determined by the TDR FP/mts probes appeared at given tested heights after 3, 5, 10 and 20 days, respectively.

In case of the prototype TDR sensor, presented in this article, time duration required for the full saturation for various heights of reference level reached 2, 4, 8, and 12 days. The measurements of rigid porous materials performed by the traditional probes had more stable process and were characterized by lower values of the determined standard deviations, approx. 0.001 cm^3/cm^3. Contrary, both, the surface TDR and FD, sensors showed values of standard deviation equal to 0.005 cm^3/cm^3, respectively. However, it should be underlined that all the determined values of standard deviations were below the extended uncertainty of TDR method. The observed differences in readouts of porous material water contents were caused by the different physical properties of tested specimens, various characteristics of sensors and varies character of the performed research.

6. Conclusions

The research on the prospective application of the surface TDR proved that the time domain reflectometry technique can be successfully utilized for noninvasive determination of moisture of rigid porous materials. Construction of the presented sensor enables to avoid the limitation of the traditional invasive probes, previously utilized only in soil science, and to extend the technology potential to other branches, mainly civil engineering. A thorough analysis of the obtained results enabled formulation of the following conclusions:

(1) For proper recalculation of reflectometric moisture readouts, the noninvasive, surface TDR sensors require individual calibration.

(2) Due to influence of polyoxymethylene cover of the sensor, apparent permittivity read by the noninvasive sensor is lower than one read by the traditional probe in relation to the same moisture level. These differences can be abolished by application of the individual calibration.

(3) Residual mean squared error (RMSE) for the calibration formula developed for the discussed sensor and material equals 0.013 cm^3/cm^3 and is smaller than found in the literature for the traditional invasive probes utilizing the standard empirical calibration formulas.

(4) Expanded uncertainty of the discussed sensor equals 0.01 cm^3/cm^3 in the most of the range of material moisture which is lower value than found in the literature for the invasive sensors utilizing the traditional empirical calibration formulas.

(5) Expanded uncertainty of the tested sensor is higher at nearly dry and nearly saturated states of the measured material.

(6) In the range of high moisture values, water content readouts by the TDR surface sensor were higher than those acquired by the capacitive sensor.

(7) In the range of average and low moisture values, water content readouts by the TDR surface sensor were lower than those acquired by the capacitive sensor.

(8) During the comparison of the indirect, electric estimation of moisture using noninvasive TDR and FD sensors with the gravimetric evaluation it was noticed that the TDR readouts were underestimated for 4.4% and the FD readouts were overestimated for 12.6%.

(9) Comparing the maximal standard deviations in both tests using electric techniques of moisture detection it was noted, that capacitive sensors are characterized by greater values of this parameter.

Author Contributions: Z.S. designed, manufactured the sensor and developed the methodology of measurement; Z.S., D.B-H. and G.Ł. conducted the laboratory investigations; Z.S., M.K.W., G.Ł. and D.M. analyzed and discussed the results; Z.S. wrote the original draft of the manuscript; M.K.W. provided language corrections; Z.S. and G.Ł. provided manuscript formatting; All authors of the article provided substantive comments.

Funding: This work was financially supported by Ministry of Science and Higher Education in Poland, within the statutory research of particular scientific units.

Conflicts of Interest: The authors declare no conflict of interest.

References

1. Topp, G.C.; Ferré, T.P.A. Time-Domain Reflectometry. In *Encyclopedia of Soils in the Environment*; Hillel, D., Ed.; Elsevier: Amsterdam, The Netherlands, 2005; pp. 174–181. ISBN 978-0-12-348530-4.

2. Rubio-Celorio, M.; Garcia-Gil, N.; Gou, P.; Arnau, J.; Fulladosa, E. Effect of temperature, high pressure and freezing/thawing of dry-cured ham slices on dielectric time domain reflectometry response. *Meat Sci.* **2015**, *100*, 91–96. [CrossRef] [PubMed]

3. Nasraoui, M.; Nowik, W.; Lubelli, B. A comparative study of hygroscopic moisture content, electrical conductivity and ion chromatography for salt assessment in plasters of historical buildings. *Constr. Build. Mater.* **2009**, *23*, 1731–1735. [CrossRef]

4. Sobczuk, H.; Plagge, R. *Time Domain Reflectometry Method in Environmental Measurements*; Polska Akademia Nauk. Komitet Inzynierii Srodowiska: Lublin, Poland, 2007; Volume 39, pp. 1–114. ISBN 83-89293-51-X.

5. Malicki, M.A.; Campbell, E.C.; Hanks, R.J. Investigations on power factor of the soil electrical impedance as related to moisture, salinity and bulk density. *Irrig. Sci.* **1989**, *10*, 55–62. [CrossRef]

6. Topp, G.C.; Davis, J.L.; Annan, A.P. Electromagnetic determination of soil water content: Measurements in coaxial transmission lines. *Water Resour. Res.* **1980**, *16*, 574–582. [CrossRef]

7. Skierucha, W.; Wilczek, A.; Alokhina, O. Calibration of a TDR probe for low soil water content measurements. *Sens. Actuators A Phys.* **2008**, *147*, 544–552. [CrossRef]

8. Topp, G.C.; Davis, J.L.; Annan, P. Electromagnetic determination of soil water content using TDR: I. Applications to wetting fronts and steep gradients. *Soil Sci. Soc. Am. J.* **1982**, *46*, 672–678. [CrossRef]

9. Malicki, M.A.; Skierucha, W.M. A manually controlled TDR soil moisture meter operating with 300 ps rise-time needle pulse. *Irrig. Sci.* **1989**, *10*, 153–163. [CrossRef]

10. Zegelin, S.J.; White, I.; Jenkins, D.J. Improved field probe for soil water content and electrical conductivity measurement using time domain reflectometry. *Water Resour. Res.* **1989**, *25*, 2367–2376. [CrossRef]

11. Černý, R. Time-domain reflectometry method and its application for measuring moisture content in porous materials: A review. *Measurement* **2009**, *42*, 329–336. [CrossRef]

12. Blonquist, J.M.; Jones, S.B.; Robinson, D.A. A time domain transmission sensor with TDR performance characteristics. *J. Hydrol.* **2005**, *314*, 235–245. [CrossRef]

13. Noborio, K. Mesurement of soil water content and electrical conductivity by time domain reflectometry: A review. *Comput. Electron. Agric.* **2001**, *31*, 213–237. [CrossRef]

14. Selker, J.S.; Graff, L.; Steenhuis, T. Noninvasive time domain reflectometry moisture measurement probe. *Soil Sci. Soc. Am. J.* **1993**, *57*, 934–936. [CrossRef]

15. Perrson, M.; Berndtsson, R. Noninvasive water content and electrical conductivity laboratory measurements using time domain reflectometry. *Soil Sci. Soc. Am. J.* **1998**, *62*, 1471–1476. [CrossRef]

16. Malicki, M.A.; Plagge, R.; Roth, C.H. Improving the calibration of dielectric TDR soil moisture determination taking into account the solid soil. *Eur. J. Soil Sci.* **1996**, *47*, 357–366. [CrossRef]

17. Sobczuk, H.; Suchorab, Z. Calibration of TDR instruments for moisture measurement of serated concrete. In *Monitoring and Modelling the Properties of Soil as Porous Medium*; Skierucha, W., Walczak, T., Eds.; Institute of Agrophysics, Polish Academy of Sciences: Lublin, Poland, 2005; pp. 156–165. ISBN 9788387385958.

18. Lee, J.; Horton, R.; Noborio, K.; Jaynes, D.B. Characterization of preferential flow in undisturbed, structured soil columns using a vertical TDR probe. *J Contam. Hydrol.* **2001**, *51*, 131–144. [CrossRef]

19. Lins, Y.; Schanz, T.; Fredlund, D.G. Modified pressure plate apparatus and column testing device for measuring SWCC of sand. *Geotech. Test J.* **2009**, *32*, 450–464. [CrossRef]

20. Skierucha, W.; Wilczek, A.; Szypłowska, A.; Sławiński, C.; Lamorski, K. A TDR-based soil moisture monitoring system with simultaneous measurement of soil temperature and electrical conductivity. *Sensors* **2012**, *12*, 13545–13566. [CrossRef] [PubMed]

21. Davis, J.L.; Annan, A.P. Ground-penetrating radar for high-resolution mapping of soil and rock stratigraphy. *Geophys. Prospect.* **1989**, *37*, 551–551. [CrossRef]

22. Skierucha, W.; Malicki, M.A. *TDR Method for the Measurement of Water Content and Salinity of Porous Media*; Institute of Agrophysics, Polish Academy of Sciences: Lublin, Poland, 2004.

23. Wilczek, A.; Szypłowska, A.; Kafarski, M.; Skierucha, W. A Time-Domain Reflectometry Method with Variable Needle Pulse Width for Measuring the Dielectric Properties of Materials. *Sensors* **2016**, *16*, 191. [CrossRef] [PubMed]

24. Jones, S.B.; Wraith, J.M.; Or, D. Time domain reflectometry measurement principles and applications. *Hydrol. Proc.* **2002**, *16*, 141–153. [CrossRef]

25. De Loor, G.P. Dielectric properties of heterogeneous mixtures containing water. *J. Microw. Power* **1968**, *3*, 67–73. [CrossRef]

26. Tinga, W.R.; Voss, W.A.G.; Blossey, D.F. Generalized approach to multiphase dielectric mixture theorie. *J. Appl. Phys.* **1973**, *44*, 3897–3902. [CrossRef]

27. Dobson, M.C.; Ulaby, F.T.; Hallikainen, M.T.; El-Rayes, M.A. Microwave dielectric behavior of wet soil. Part 2: Dielectric mixing models. *IEEE Trans. Geosci. Remote Sens.* **1985**, *23*, 35–46. [CrossRef]

28. Noborio, K.; Horton, R.; Tan, C.S. Time Domain Reflectometry probe for simultaneous measurement of soil matric potential and water content. *Soil Sci. Soc. Am. J.* **1999**, *63*, 1500–1505. [CrossRef]

29. O'Connor, K.M.; Dowding, C.H. *Geomeasurements by Pulsing TDR Cables and Probes*; CRC Press: Boca Raton, FL, USA, 1999; ISBN 9780849305863.

30. Moret, D.; Lopez, M.V.; Arrue, J.L. TDR application for automated water level measurement from Mariotte reservoirs in tension disc infiltrometers. *J. Hydrol.* **2004**, *297*, 229–235. [CrossRef]

31. Topp, G.C.; Reynolds, W.D. Time Domain Reflectometry: A seminar technique for measuring mass and energy in soil. *Soil Tillage Res.* **1998**, *47*, 125–132. [CrossRef]

32. Jones, S.B.; Or, D. Modeled effects on permittivity measurements of water content in high surface area porous media. *Physica B* **2003**, *338*, 284–290. [CrossRef]

33. Skierucha, W. Accuracy of Soil Moisture Measurement by TDR Technique. *Int. Agrophys.* **2000**, *14*, 417–426.

34. Rayleigh, L. On the influence of obstacles arranged in rectangular order upon the properties of the medium. *Philos. Mag.* **1892**, *34*, 481–502. [CrossRef]

35. Maxwell Garnett, J.C. *Colours in Metal Gases and Metal Films*; Transactions of the Royal Society: London, UK, 1904; Volume 203, pp. 385–420.

36. Polder, D.; van Santen, J.H. The effective permeability of mixtures of solids. *Physica* **1946**, *12*, 257–271. [CrossRef]

37. Roth, K.; Schulin, R.; Flühler, H.; Attinger, W. Calibration of time domain reflectometry for water content measurement using a composite dielectric approach. *Water Resour. Res.* **1990**, *26*, 2267–2273. [CrossRef]

38. Whalley, W.R. Consideration on the use of Time Domain Reflectometry (TDR) for measuring soil water content. *Eur. J. Soil Sci.* **1993**, *44*, 1–9. [CrossRef]

39. Schapp, M.G.; de Lange, L.; Heimovara, T.J. TDR calibration of organic forest floor media. *Soil Technol.* **1996**, *11*, 205–217. [CrossRef]

40. Skierucha, W. *Wpływ Temperatury Na Pomiar Wilgotności Gleby Metodą Reflektometryczną*; Acta Agrophysica, Rozprawy i Monografie, Polska Akademia Nauk: Lublin, Polska, 2005. (In Polish)

41. Quinones, H.; Ruelle, P. Operative calibration methodology of a TDR sensor for soil moisture monitoring under irrigated crops. *Subsurf. Sens. Technol. Appl.* **2001**, *2*, 31–45. [CrossRef]

42. Udawatta, R.P.; Anderson, S.H.; Motavalli, P.P.; Garrett, H.E. Calibration of a water content reflectometer and soil water dynamics for an agroforestry practice. *Agrofor. Syst.* **2011**, *82*, 61–75. [CrossRef]

43. Mastrorilli, M.; Katerji, N.; Rana, G.; Nouna, B.B. Daily actual evapotranspiration measured with TDR technique in Mediterranean conditions. *Agric. For. Meteorol.* **1998**, *90*, 81–89. [CrossRef]

44. Ren, T.; Noborio, K.; Horton, R. Measuring soil water content, electrical conductivity, and thermal properties with a thermo-time domain reflectometry probe. *Soil. Sci. Soc. Am. J.* **1999**, *63*, 450–457. [CrossRef]

45. Soncela, R.; Sampaio, S.; Vilas Boas, M.A.; Tavares, M.H.F.; Smanhotto, A. Construction and calibration of TDR probes for volumetric water content estimation in a Distroferric Red Latosol. *Eng. Agríc.* **2013**, *33*, 919–928. [CrossRef]

46. Ju, Z.; Liu, X.; Ren, T.; Hu, C. Measuring Soil Water Content with Time Domain Reflectometry: An Improved Calibration Considering Soil Bulk Density. *Soil Sci.* **2010**, *175*, 469–473. [CrossRef]

47. Hansen, M.H. TDR measurement of moisture content in aerated concrete. In Proceedings of the 6th Symposium on Building Physics, Trondheim, Norway, 17–19 June 2002.

48. Suchorab, Z.; Widomski, M.; Łagód, G.; Sobczuk, H. Capillary rise phenomenon in aerated concrete. Monitoring and simulations. *Proc. ECOpole* **2010**, *4*, 285–290.

49. Suchorab, Z.; Barnat-Hunek, D.; Franus, M.; Łagód, G. Mechanical and Physical Properties of Hydrophobized Lightweight Aggregate Concrete with Sewage Sludge. *Materials* **2016**, *9*, 317. [CrossRef] [PubMed]

50. Pavlík, Z.; Fiala, L.; Černý, R. Determination of Moisture Content of Hygroscopic Building Materials Using Time Domain Reflectometry. *J. Appl. Sci.* **2008**, *8*, 1732–1737. [CrossRef]

51. Wraith, J.M.; Robinson, D.A.; Jones, S.B.; Long, D.S. Spatially characterizing apparent electrical conductivity and water content of surface soils with time domain reflectometry. *Comput. Electron. Agric.* **2005**, *46*, 239–261. [CrossRef]

52. Ito, Y.; Chikushi, J.; Miyamoto, H. Multi-TDR probe designer for measuring soil moisture distribution near the soil surface. In Proceedings of the 19th World Congress of Soil Sciences, Brisbane, Australia: Soil Solutions for a Changing World, Brisbane, Australia, 1–6 August 2010.

53. Choi, C.; Song, M.; Kim, D.; Yu, X. A New Non-Destructive TDR System Combined with a Piezoelectric Stack for Measuring Properties of Geomaterials. *Materials* **2016**, *9*, 439. [CrossRef] [PubMed]

54. Sobczuk, H. Sonda Do Pomiaru Wilgotności Ośrodków Porowatych, Zwłaszcza Materiałów Budowlanych. Available online: https://rejestr.io/patenty/212837 (accessed on 13 November 2018).

55. Sobczuk, H.; Suchorab, Z. Sonda Do Pomiaru Wilgotności Przegród Budowlanych, Zwłaszcza O Chropowatych Powierzchniach. Available online: https://rejestr.io/patenty/225640 (accessed on 13 November 2018).

56. Sobczuk, H.; Suchorab, Z. Sonda Do Pomiaru Wilgotności, Zwłaszcza Elementów O Powierzchniach Wypukłych. Available online: https://rejestr.io/patenty/225641 (accessed on 13 November 2018).

57. Sobczuk, H.; Suchorab, Z. Sonda Do Pomiaru Wilgotności, Zwłaszcza Elementów O Zakrzywionych Powierzchniach. Available online: https://rejestr.io/patenty/225639 (accessed on 13 November 2018).

58. Knight, J.H. Sensitivity of time domain reflectometry measurements to lateral variations in soil water content. *Water Resour. Res.* **1992**, *28*, 2345–2352. [CrossRef]

59. Suchorab, Z.; Sobczuk, H.; Cerny, R.; Pavlik, Z.; De Miguel, R.S. Sensitivity range determination of surface TDR probes. *Environ. Prot. Eng.* **2009**, *35*, 179–189.

60. Majerek, D.; Widomski, M.; Garbacz, M.; Suchorab, Z. Estimation of the measurement uncertainty of humidity using a TDR probe. *AIP Conf. Proc.* **2018**, *1988*, 020027. [CrossRef]

61. JCGM 100:2008, GUM 1995 with Minor Corrections, Evaluation of Measurement Data—Guide to the Expression of Uncertainty in Measurement. Available online: http://www.bipm.org/utils/common/documents/jcgm/JCGM_100_2008_E.pdf (accessed on 24 September 2018).

62. Wu, S.Y.; Zhou, Q.Y.; Wang, G.; Yang, L.; Ling, C.P. The relationship between electrical capacitance-based dielectric constant and soil water content. *Environ. Earth Sci.* **2010**, *62*, 999–1011. [CrossRef]

63. Topp, G.C.; Davis, J.L.; Bailey, W.G.; Zebchuk, W.D. The measurement of soil water content using a portable TDR hand probe. *Can. J. Soil. Sci.* **1984**, *64*, 313–321. [CrossRef]

64. Amato, M.; Ritchie, J.T. Small spatial scale soi water content measurement with time-domain reflectometry. *Soil Sci. Soc. Am. J.* **1995**, *59*, 325–329. [CrossRef]

sensors

MDPI

Article

A Novel Method and an Equipment for Generating the Standard Moisture in Gas Flowing through a Pipe

Yusuke Tsukahara *, Osamu Hirayama, Nobuo Takeda, Toru Oizumi, Hideyuki Fukushi, Nagisa Sato, Toshihiro Tsuji, Kazushi Yamanaka and Shingo Akao

Ball Wave Inc., T-Biz 503, 6-6-40, Aza Aoba, Aramaki, Aoba, Sendai, Miyagi 980-8579, Japan; oflathill@gmail.com (O.H.); takeda@ballwave.jp (N.T.); oizumi@ballwave.jp (T.O.); fukushi@ballwave.jp (H.F.); n.is2v.w@gmail.com (N.S.); t-tsuji@material.tohoku.ac.jp (T.T.); yamanaka@ballwave.jp (K.Y.); akao@ballwave.jp (S.A.)
* Correspondence: tsukahara@ballwave.jp; Tel.: +81-80-6768-5136

Received: 3 September 2018; Accepted: 8 October 2018; Published: 13 October 2018

Abstract: When inert gas containing water molecules flows into a metal pipe, the water molecules cannot exit instantaneously from the outlet of the pipe but are captured at adsorption sites on the inner surface of the pipe until most of the sites are occupied. A theoretical model and a subsequent experiment in this article show that the delay time depends on the amount of moisture level; the higher the moisture-level, the shorter the delay time. Based on the result, we propose a new method and its implementation to the validation of a standard moisture generation to be used in the field measurement such as in factories and pipe lines.

Keywords: Trace moisture; ball SAW sensor; surface acoustic wave; permeation tube

1. Introduction

The measurement and control of trace moisture in gaseous materials are an important step for the quality enhancement in manufacturing semiconductors and light emitting displays [1]. There are various technologies for trace moisture measurement including aluminum oxide sensors [2], tunable laser diodes [3], and the cavity ring down spectroscopy [4]. Recently, the present authors developed a new technology called ball surface acoustic wave (SAW) moisture sensors [5]. It covers a wide range of moisture level, from a few ppbV to hundreds ppmV, and the most prominent characteristic is its quick response to a sudden variation in the moisture level [6].

For any methods of moisture measurement, calibration is inevitable, and the calibration of a particular sensor should be traceable to an international standard [7]. It is also important to periodically validate the sensor accuracy against the calibrated values while the sensor is running in the field of measurement such as in factories and in pipe lines [8]. To calibrate and to validate the moisture sensors, the accurate generation of moisture at certain predetermined values is crucial. Different methods have been proposed and implemented including the diffusion tube method [9], NPL method [10] and a method using permeation tubes [11].

Of these methods, the one which uses the permeation tube is suitable to be implemented in an equipment for the on-site validation because of its small volume. The permeation tube is made of a polymer tube with a certain diameter and a length containing liquid water in it. A tightly sealed cell containing the permeation tube is connected to a pipe line, and the gas flows into and out of the cell at a regulated constant flow rate. The polymeric material is permeable to water molecules, and it therefore dispenses the water molecules at a constant rate when the temperature and pressure are maintained constant. The amount of moisture generated by the permeation tube is controlled by changing the temperature. As a result, the amount of moisture in the output gas from the cell is a sum of the original moisture in the input gas and that generated by the permeation tube. Therefore, the output gas can be used as a standard moisture for the validation when we can guarantee that the amount of moisture contained in the input gas is small enough. To this end, dryers containing desiccants such as silica particles are used. However, the dryer stops absorbing the water molecules when it becomes saturated, therefore we need to know if the dryer is functioning properly.

In this paper, we propose a new method and its implementation to guarantee that a dryer connected to the inlet of the cell containing the permeation tube is working properly. By using the new method, the validation of the dryer output gas using the ball SAW moisture sensor is easily achieved in the field measurement such as in factories and pipe lines.

2. Moisture in a Gas Flowing through a Pipe

In the first place, we analyze the behavior of water molecules in a gas flowing through a metal pipe. The water molecules are readily adsorbed to the inner surface of pipes, cylinders, and chambers when a gas containing moisture flows through them. The water molecules are one of the contaminants seriously affecting the quality of products processed using these pipes, cylinders, and chambers [12]. It has been shown that a smaller amount of water molecule is adsorbed on a smoother inner surface of a metal pipe [13]. Recently, the present authors showed [14] that a quantitative analysis of correlation was possible between the degree of surface treatment such as electrochemical buffing (ECB) and electropolishing (EP), and the amount of water adsorption when a ball SAW moisture sensor monitored the time-dependence of moisture in a gas passing through a metal pipe only 10 cm long. This was made possible because the ball SAW sensor had a quick response time within a few seconds. In the following, we propose a theoretical model to describe the time-dependence of the moisture level in an infinitesimally small volume of an inert gas that is flowing through a pipe.

3. Theoretical Analysis

There have been theoretical and experimental studies on the behavior of molecules contained in a carrier gas passing through a column in a gas chromatograph [15,16]. However, the strength of the interaction of those molecules with the inner surface of the column is basically a linear function of the number of the molecules, and the secondary nonlinear effect was taken into account for the detail analysis of deviation from the linear model. This correctly reflected the most prominent feature of gas chromatography that the retention time is independent of the number of molecules of interest. In contrast, the adsorption to, and desorption from, the metal surface of water molecules seem to be fundamentally nonlinear function of the moisture, as shown in the following. A detail of the model is found in the Appendix A.

Let us assume that an inert gas is flowing at constant flow rate f (m^3 s^{-1}) through a pipe with a length L (m) and an inner diameter d (m) as depicted in Figure 1. The surface density of adsorption sites is s (mol m^{-2}), and an adsorption ratio to the sites, or the ratio of the number of adsorption sites occupied by the water molecules to the total number of adsorption sites, is r. The normalized moisture in the gas is W, that is

$$W = (w \times L3)/(s \times L2) \tag{1}$$

where w is the moisture measured in $\left(\text{mol m}^{-3} \right)$. A set of two normalized dimensionless equations is given as follows where a and b are the only adjustable parameters, as shown in the Appendix A (Equations (A9) and (A10)).

$$\frac{\partial r}{\partial \tau} = -a \times r + b \times (1-r) \times W \quad \frac{\partial W}{\partial \tau} + \frac{\partial W}{\partial \xi} = g \times a \times r - g \times b \times (1-r) \times W \tag{2}$$

where τ and ξ are normalized time and space coordinate, respectively, defined by Equations (A11) and (A12) in the Appendix A.

$$\tau = \frac{4 \times t \times f}{\pi \times L \times d^2} \tag{3}$$

$$\zeta = \frac{x}{L} \tag{4}$$

where t and x are time and space coordinates measured in (s) and (m), respectively. A computer program was developed in Fortran language to numerically solve the equations.

To simulate the experiments in [14], we set the values of parameters as follows.

f: 0.1 L/min

L: 10 cm

d: 4.35 mm

w_0: 1 ppbV

w_1: 1 ppmV

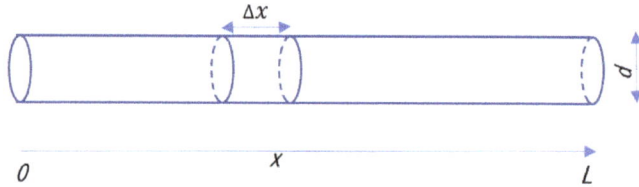

Figure 1. Inert gas containing moisture flows through a pipe with a length L and an inner diameter d.

Assuming $a = 1$ and $b = 1$ (for simplicity) and adjusting s, we obtain the time-dependence of moisture measured at the outlet of the pipe for ECB and EP tubes, respectively, as shown in Figure 2. The leading edges of time-dependence of moisture for ECB and EP tubes reasonably match the experimental values of 15 (s) and 40 (s), respectively, in Reference [14].

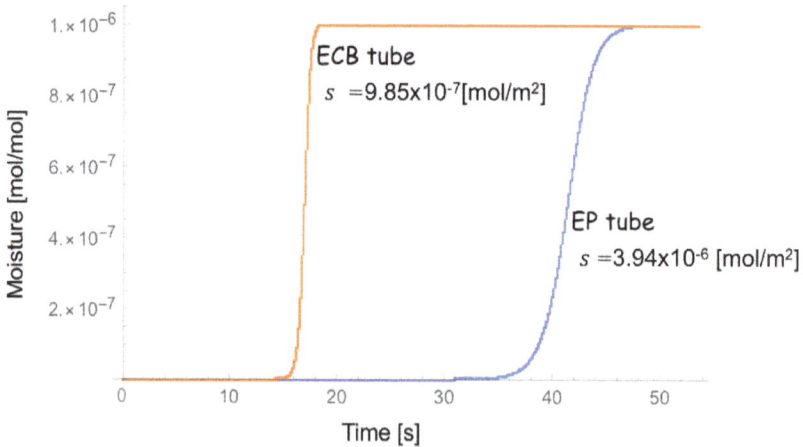

Figure 2. Calculated time-dependence of moisture at the outlet of the pipe for electrochemical buffing (ECB) tube (orange) and electropolishimg (EP) tube (blue), respectively.

Now we simulate the behavior of moisture in a setting depicted in Figure 3, where an inert gas with unknown moisture passes through a dryer and then flows into a cell containing a permeation tube. The gas coming out of the cell goes into an EP tube with $L = 16$ (cm). We chose the EP tube because it would cause a long delay time, which was easy to detect. We assume that the temperature of the permeation tube is controlled so that it generates 1 or 5 ppmV of moisture. Then, we solve the equations for the moisture in the EP tube with a set of different initial conditions which simulates the uncontrollable variation of the moisture coming out of the dryer.

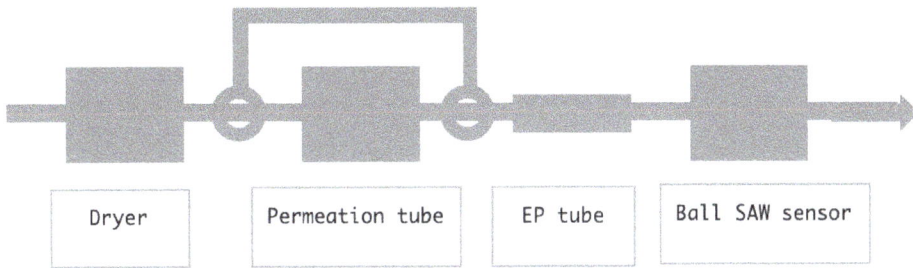

Figure 3. Inert gas with unknown moisture passes through a dryer, then flows into a cell containing a permeation tube and then into a cell for the ball surface acoustic wave (SAW) moisture sensor.

Figure 4 shows the calculated time-dependence of moisture at the outlet of the EP tube for the different set of initial conditions at the inlet of the EP tube. This simulates the situation where the output of the dryer contains the moisture of 0.05, 0.20, 0.50, and 1.00 ppmV, respectively, and then the permeation tube adds 1 ppmV of moisture. In Figure 4, we can see that the dryer's performance can be evaluated by measuring the delay time between the onset of gas flow and the leading edge of the moisture change at the outlet of the EP tube.

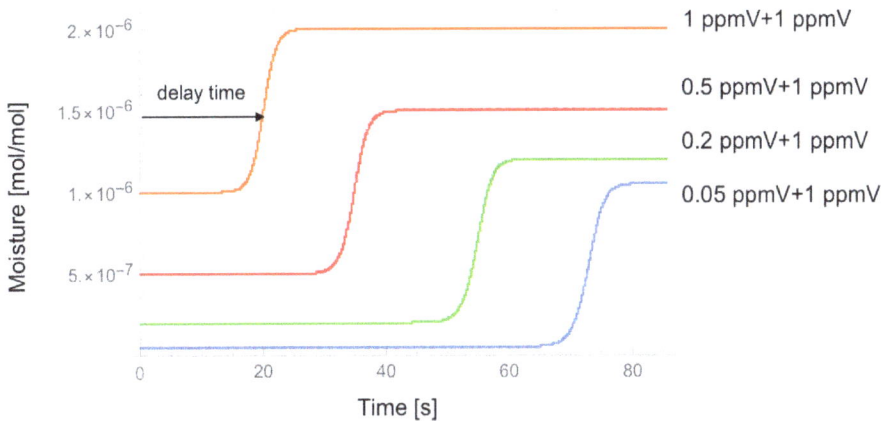

Figure 4. Calculated time-dependence of moisture at the outlet of the EP tube for the different set of initial moisture conditions at the inlet of the EP tube.

Figure 5 shows the similar analysis where the output of the dryer contains the moisture of 0.06 ppmV, 0.2 ppmV, and 0.5 ppmV, respectively, and then the permeation tube adds 5 ppmV moisture. It is still valid that the dryer's performance can be evaluated by measuring the delay time between the onset of gas flow and the leading edge of the moisture change at the outlet of the EP tube, though the time difference is smaller for the larger moisture.

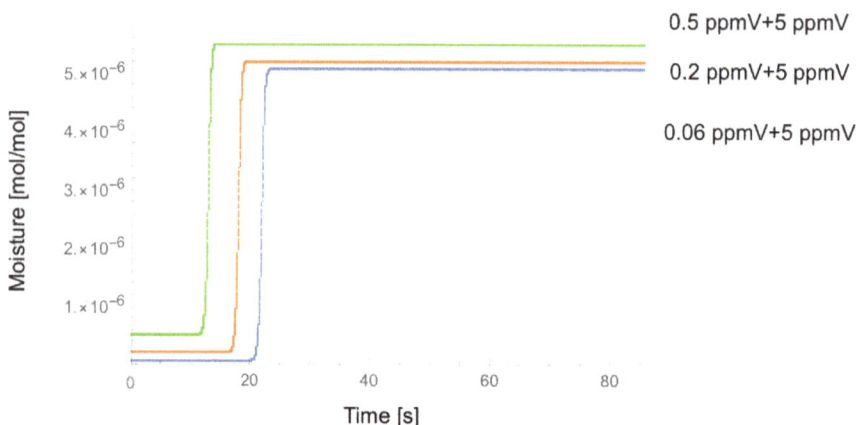

Figure 5. Calculated time-dependence of moisture at the outlet of the EP tube where the output of the dryer contains the moisture of 0.06, 0.20 and 0.50 ppmV, respectively, and then the permeation tube adds 5 ppmV of moisture.

4. Result

Figure 6 shows the experimental setup to validate the theoretical prediction. Nitrogen gas with controlled values of moisture is fed into a cell containing a permeation tube, and then goes into a 10 (cm) long EP tube, depicted as Delay. The gas coming out of the EP tube flows through a metal-mesh filter for removal of particles before reaching the measurement cell of ball SAW moisture sensor, depicted as FT. MFC1~MFC3 and MFC6 are mass flow controllers. Fine Purer is a dryer marketed by Osaka Gas Liquid Co., Ltd. in Osaka, Japan. Its specification declares that the gas coming out of it contains "< 1 nmol/mol for H_2O", which is less than 1 ppbV. Diffusion tube is providing the water molecules into the piping system. The "CRDS" block represents a CRDS Trace Gas Analyzer, HALO 3 H_2O by Tiger Optics in Pennsylvania, USA, used as a reference. Its specification declares that the detection range and the low detection limit for H_2O in nitrogen are 0–20 ppmV and 0.6 ppbV, respectively. In the numerical calculation in the previous section, the effect of the metal-mesh filter was taken into account by assuming the 16 (cm) long EP tube. The ball SAW sensor was driven with an electric pulse containing two different frequency components, namely 80 and 240 MHz. The two frequency components of the output signal were subtracted to compensate for the temperature dependence of the sensor, and then converted to the values of moisture content.

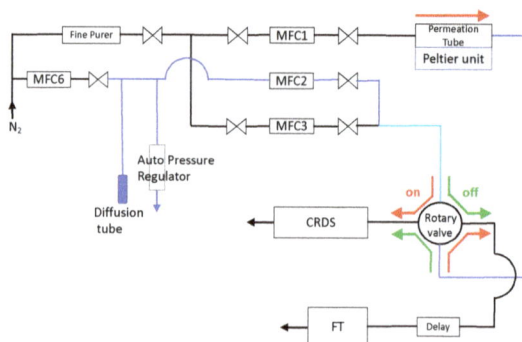

Figure 6. Experimental setup. MFC's: mass flow controllers, Fine Pure: a dryer, CRDS: a CRDS Trace Gas Analyzer, Delay: a 10 (cm) long EP tube, FT: a ball SAW moisture sensor.

Figure 7 shows the measured signals of the ball SAW moisture sensor for four different conditions: (a) 0.05, (b) 0.2, (c) 0.5, and (d) 1.0 ppmV of background moisture, each time mixed with 1 ppmV from permeation tube. The vertical axis is the normalized value of moisture measured by the ball SAW sensor because the absolute value is not calibrated yet. The experiment was repeated for four times with each condition.

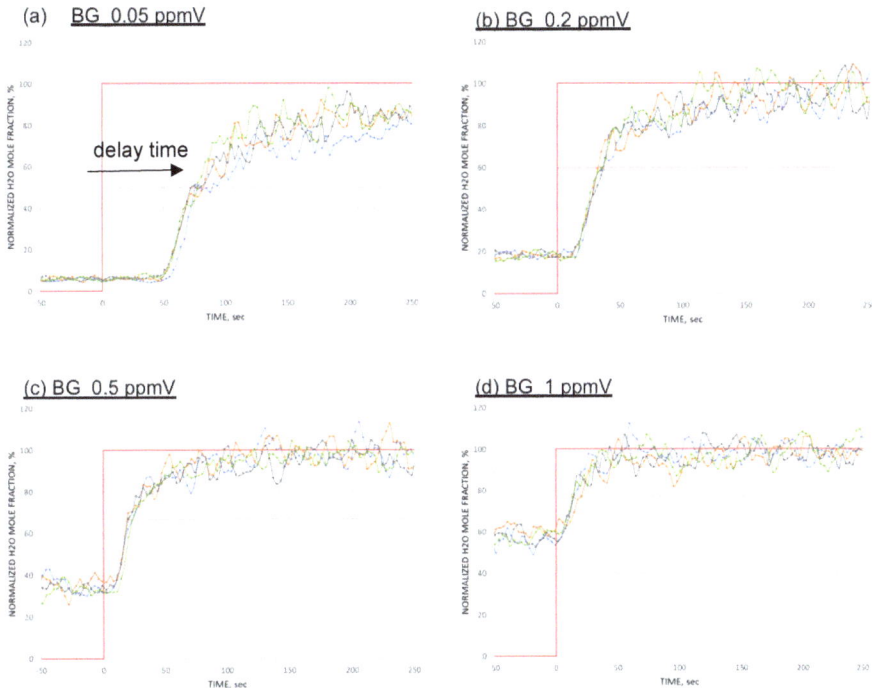

Figure 7. Measured signals of ball SAW moisture sensor for four different conditions: (a) 0.05, (b) 0.2, (c) 0.5, and (d) 1.0 ppmV of background moisture, each time mixed with 1 ppmV from permeation tube. The vertical axis is for normalized values. BG stands for back ground moisture.

Figure 8 show the measured signals for three different conditions: (a) 0.06, (b) 0.2, and (c) 0.5 ppmV of background moisture, each time mixed with 5 ppmV from permeation tube. It should be noted that the delay time between the onset and the leading edge of the moisture change depends on the background moisture as predicted by the theoretical simulation.

More quantitatively, the theoretical and experimental delay time is plotted in Figures 9 and 10. The theoretical and experimental values do not exactly match, but the trend is reproduced correctly and the smaller the background moisture, the larger the delay time.

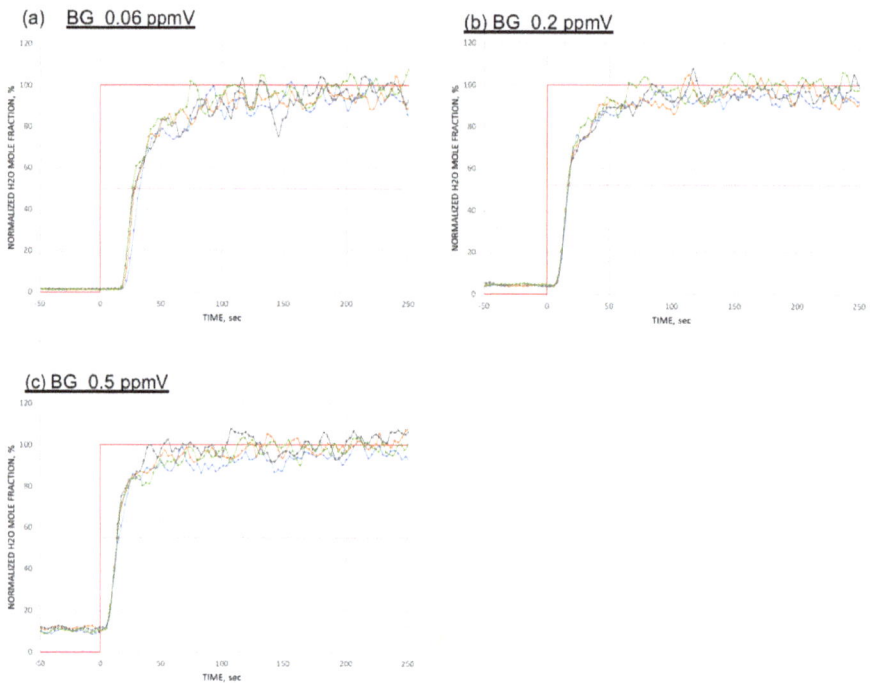

Figure 8. Measured signals for (**a**) 0.06, (**b**) 0.2, and (**c**) 0.5 ppmV of background moisture, each time mixed with 5 ppmV from permeation tube. The vertical axis is for normalized values. BG stands for back ground moisture.

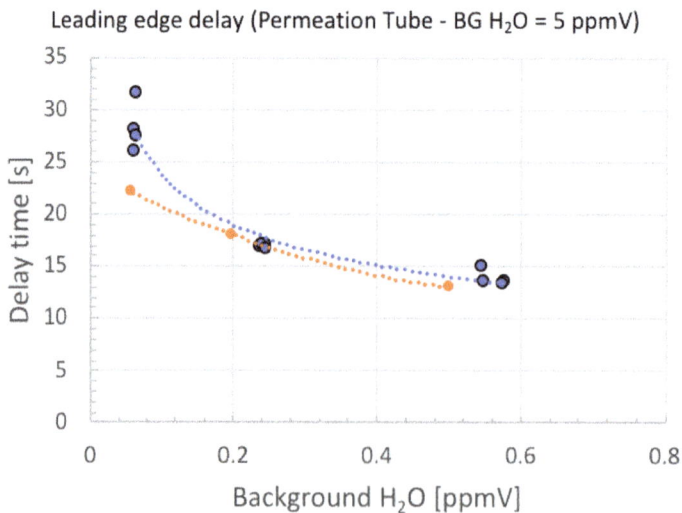

Figure 9. Theoretical (orange) and experimental (blue) delay time plotted as a function of background moisture when the permeation tube generates 5 ppmV moisture. Dotted lines are fitted curves.

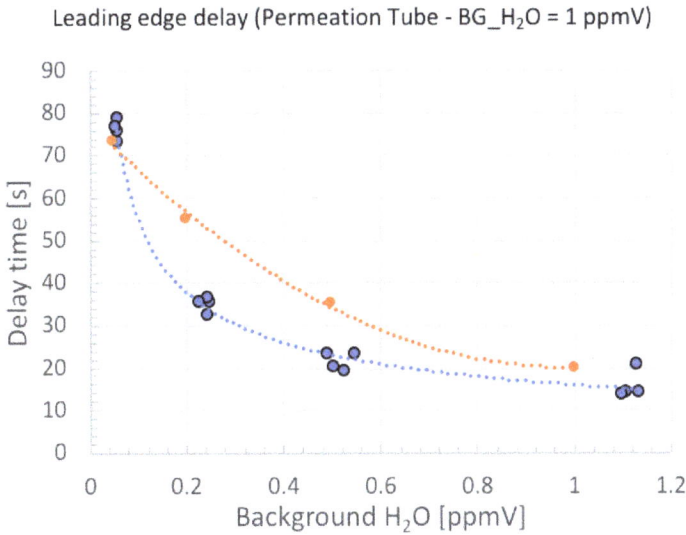

Figure 10. Theoretical (orange) and experimental (blue) delay time plotted as a function of background moisture when the permeation tube generates 1 ppmV moisture. Dotted lines are fitted curves.

5. Discussion

In this article, it is established that choice of a metal pipe with proper inner surface treatment in combination with a ball SAW moisture sensor can be used for the evaluation of a background moisture in a gas coming out of a dryer. This is an original novel design of a standard moisture generator with a permeation tube, particularly suitable for the validation of trace moisture sensors in the field measurement, such as in factories and in pipelines.

The experimental data in Figure 9 shows that the background moisture of 0.5 ppmV added to the 5 ppmV standard moisture gave rise to the time delay of −50%. In Figure 10, the background moisture of 0.1 ppmV added to the 1 ppmV standard moisture gave rise to the time delay of −27%. Therefore, we conclude that by measuring the delay time, we can easily distinguish the uncontrollable background moisture at less than 10%. This is a unique way of guaranteeing the accuracy of the standard moisture for the validation of ball SAW moisture sensors.

Author Contributions: Conceptualization and Investigation, Y.T.; Methodology, N.T.; Software, O.H.; Validation, T.T., K.Y., H.F., N.S. and T.O.; Formal Analysis, Y.T.; Writing-Original Draft Preparation, Y.T.; Writing-Review & Editing, Y.T.; Visualization, T.O. and H.F.; Project Administration, S.A.

Conflicts of Interest: The authors declare no conflict of interest.

Appendix A

Let us assume that an ideal gas containing water molecules flows through a pipe with a length L [m] and an inner diameter d [m] at a constant flow rate f [m^3 s^{-1}]. We assume that the flow velocity is uniform over the entire cross-section of the pipe for simplicity. Then, the flow velocity v is

$$v = \frac{4 \times f}{\pi \times d^2} \tag{A1}$$

We propose that there are microscopic sites on the inner surface of the metal pipe with a surface density s $\left[\text{mol m}^{-2}\right]$ where water molecules can be adsorbed. The water molecules attach to and detach from these sites at each instant, and on average there are $s \times r$ sites with water molecules adsorbed per unit area, where r is an adsorption ratio.

In the experiment [14], the nitrogen gas with a constant moisture w_0 $\left(\text{mol m}^{-3}\right)$ flows into the pipe from the inlet ($x = 0$). The moisture at the outlet ($x = L$) eventually becomes equal to that at the inlet after a sufficiently long time. Then, the moisture of the gas flowing into the inlet is suddenly changed to another constant value w_1 $\left(\text{mol m}^{-3}\right)$ at time $t = 0$, and the time dependence of the moisture at the outlet ($w(t, x = L)$) is monitored.

The amount of water molecules that detach from the unit area of the inner surface and enter into the carrier gas is proportional to the amount of adsorbed water molecules on the unit area, $s \times r$. Thus, introducing a desorption coefficient k_d, it is $s \times r \times k_d$. On the other hand, the amount of water molecules adsorbed on the unit area of the inner surface is proportional to the product of the number of vacant sites, $s \times (1 - r)$, and the number of water molecules in the carrier gas or moisture, w. Thus, it is $s \times (1 - r) \times k_a \times w$ where k_a is an adsorption coefficient.

Now let us take a volume with a length Δx at a position $x = x_0$ along the length of the pipe. The amount of water molecules contained in the volume increases due to the incoming flow from the upstream and decreases due to the outgoing flow to the downstream, and the net increase due to the flow during a time interval Δt is their difference,

$$-\frac{\partial w(t, x)}{\partial x} \times v \times \left(\frac{d}{2}\right)^2 \times \pi \times \Delta x \times \Delta t \tag{A2}$$

In this time interval, a part of the water molecules in the gas flow are adsorbed on the metal surface and a part of the water molecules on the surface detach from the surface, and the net amount of adsorbed molecules is,

$$[s\, k_d\, r(t, x) - s\, k_a\, (1 - r(t, x))\, w(t, x)] \times \pi \times d \times \Delta x \times \Delta t \tag{A3}$$

The diffusion of water molecules may occur when there is a difference of moisture along the length of the pipe, but we ignore it assuming its effect is smaller than that of the Equations (A2) and (A3). We can take it into account if necessary by introducing a diffusion term

$$D\frac{\partial^2 w}{\partial x^2} \left(\frac{d}{2}\right)^2 \times \pi \times \Delta x \times \Delta t \tag{A4}$$

The net increase of water molecules during a time interval Δt in the volume is

$$\frac{\partial w}{\partial t} \left(\frac{d}{2}\right)^2 \times \pi \times \Delta x \times \Delta t \tag{A5}$$

which is equal to a sum of Expressions (A2) and (A3),

$$\tfrac{\partial w}{\partial t}\left(\tfrac{d}{2}\right)^2 \times \pi \times \Delta x \times \Delta t = -\tfrac{\partial w}{\partial x}v \cdot \left(\tfrac{d}{2}\right)^2 \times \pi \times \Delta x \times \Delta t + [s\, k_d\, r - s\, k_a\, (1 - r)\, w] \times \pi \times d \times \Delta x \times \Delta t \tag{A6}$$

Then, the net increase of the adsorbed molecules on the inner surface during a time interval Δt is

$$s\frac{\partial r}{\partial t} \times \pi \times d \times \Delta x \times \Delta t \tag{A7}$$

which is equal to the Expression (A3),

$$s\frac{\partial r}{\partial t} \times \pi \times d \times \Delta x \times \Delta t = [s\, k_d\, r - s\, k_a\, (1 - r)\, w] \times \pi \times d \times \Delta x \times \Delta t \tag{A8}$$

In a dimensionless form, the Equations (A6) and (A8) become

$$\frac{\partial r}{\partial \tau} = -a \times r + b \times (1 - r) \times W \tag{A9}$$

$$\frac{\partial W}{\partial \tau} + \frac{\partial W}{\partial \zeta} = g \times a \times r - g \times b \times (1 - r) \times W \tag{A10}$$

where

$$\tau = \frac{t \times v}{L} \tag{A11}$$

$$\zeta = \frac{x}{L} \tag{A12}$$

$$W = \frac{w \times L^3}{s \times L^2} \tag{A13}$$

$$a = \frac{L \times k_d}{v} \tag{A14}$$

$$b = \frac{s \times k_a}{v} \tag{A15}$$

$$g = \frac{4 \times L}{d} \tag{A16}$$

In the equilibrium

$$\frac{\partial W}{\partial \tau} = \frac{\partial r}{\partial \tau} = 0 \tag{A17}$$

therefore, from Equations (A9) and (A10),

$$W = constant = W_0 \tag{A18}$$

and

$$r = r_1 = \frac{1}{1 + \frac{a}{b} \times \frac{1}{W_0}} \tag{A19}$$

This means that the adsorption ratio r reaches r_1 regardless of the position along the length of the pipe x when the gas with a constant moisture, $W = constant = W_0$, flows through the pipe for a long time. Furthermore, when a very dry gas comes in, no adsorption sites are occupied,

$$r_1 \to 0 \text{ when } (W_0 \to 0) \tag{A20}$$

whereas all site will be occupied when a very wet gas comes in.

$$r_1 \to 1 \text{ when } (W_0 \to \infty) \tag{A21}$$

which is obvious.

We can obtain the temporal evolution of the system by solving the Equations (A9) and (A10) under the initial condition,

$$W(\tau = 0, \zeta) = W_0 \text{ for } (0 \le \zeta \le 1) \tag{A22}$$

$$r(\tau = 0, \zeta) = r_1 \text{ for } (0 \le \zeta \le 1) \tag{A23}$$

and the boundary condition,

$$W(\tau, \zeta = 0) = W_1 \text{ for } (0 < \tau) \tag{A24}$$

We numerically solve the equations with the values of parameters taken from the experiments [14], but the coefficients k_d and k_a are arbitrary assumed such that $a \sim 1$ and $a \sim b$. This is equivalent

Sensors **2018**, *18*, 3438

to the assumption that the contribution of adsorption and desorption is in the same order in the Equation (A9).

References

1. Funke, H.H.; Grissom, B.L.; McGrew, C.E.; Raynor, M.W. Techniques for the measurement of trace moisture in high-purity electronic specialty gases. *Rev. Sci. Instrum.* **2003**, *74*, 3909. [CrossRef]
2. Cecil, O.B. Aluminum oxide humidity sensor. U.S. Patent 3,440,372, 22 April 1969.
3. Zhou, X.; Liu, X.; Jeffries, J.B.; Hanson, R.K. Development of a sensor for temperature and water concentration in combustion gases using a single tunable diode laser. *Meas. Sci. Technol.* **2003**, *14*, 1459. [CrossRef]
4. Demtroeder, W. *Laser Spectroscopy*, 4th ed.; Springer: Berlin, Germany, 2008; p. 22.
5. Takeda, N.; Motozawa, M. Extremely Fast 1 μmol·mol^{-1} Water-Vapor Measurement by a 1 mm Diameter Spherical SAW Device. *Int. J. Thermophys.* **2012**, *33*, 1642. [CrossRef]
6. Tsuji, T.; Oizumi, T.; Takeda, N.; Akao, S.; Tsukahara, Y.; Yamanaka, K. Temperature compensation of ball surface acoustic wave sensor by two-frequency measurement using undersampling. *Jpn. J. Appl. Phys.* **2015**, *54*, 07HD13. [CrossRef]
7. Abe, H.; Yamada, K.M.T. Performance evaluation of a trace-moisture analyzer based on cavity ring-down spectroscopy: Direct comparison with the NMIJ trace-moisture standard. *Sens. Actuators A Phys.* **2011**, *165*, 230. [CrossRef]
8. McKeogh, G.; Sparages, N.J.D. 16-McKeogh-1. In Proceedings of the GAS2017, Rotterdam, The Netherlands, 13 June 2017.
9. Abe, H.; Kitano, H. Development of humidity standard in trace-moisture region: Characteristics of humidity generation of diffusion tube humidity generator. *Sens. Actuators A Phys.* **2006**, *128*, 202. [CrossRef]
10. Stevens, M.; Bell, S.A. The NPL standard humidity generator: an analysis of uncertainty by validation of individual component performance. *Meas. Sci. Technol.* **1992**, *3*, 943. [CrossRef]
11. O'Keeffe, A.E.; Ortman, G.C. Primary standards for trace gas analysis. *Anal. Chem.* **1966**, *38*, 760. [CrossRef]
12. Ohmi, T. Ultra clean processing. *Microeclectr. Eng.* **1991**, *10*, 163. [CrossRef]
13. Ohmi, T.; Nakagawa, Y.; Nakamura, M.; Ohki, A.; Koyama, T. Formation of chromium oxide on 316L austenitic stainless steel. *J. Vac. Sci. Technol. A* **1996**, *14*, 2505. [CrossRef]
14. Tsuji, T.; Akao, S.; Oizumi, T.; Takeda, N.; Tsukahara, Y.; Yamanaka, K. Moisture adsorption desorption characteristics of stainless steel tubing measured by ball surface acoustic wave trace moisture analyzer. *Jpn. J. Appl Phys.* **2017**, *56*, 07JC03. [CrossRef]
15. Jaulmes, A.; Vidal-Madjar, C.; Ladurelli, A.; Guiochon, G. Study of peak profiles in nonlinear gas chromatography. 1. Derivation of a theoretical model. *J. Phys. Chem.* **1984**, *88*, 5379. [CrossRef]
16. Jaulmes, A.; Vidal-Madjar, C.; Ladurelli, A.; Guiochon, G. Study of peak profiles in nonlinear gas chromatography. 2. Determination of the curvature of isotherms at zero surface coverage on graphitized carbon black. *J. Phys. Chem.* **1984**, *88*, 5385. [CrossRef]

sensors

MDPI

Article

Reliability Modeling for Humidity Sensors Subject to Multiple Dependent Competing Failure Processes with Self-Recovery

Jia Qi [1], Zhen Zhou [1,*], Chenchen Niu [1], Chunyu Wang [1] and Juan Wu [2]

[1] School of Measurement and Communication Engineering, Harbin University of Science and Technology, Harbin 150080, China; qjia89@hrbust.edu.cn (J.Q.); samueland@126.com (C.N.); wangchunyu230281@163.com (C.W.)

[2] College of Precision Instruments and Opto-electronics Engineering, Tianjin University, Tianjin 300072, China; taozi_xixi@163.com

* Correspondence: zhzh49@126.com; Tel.: +86-130-0986-1061 or +86-0451-8639-2318

Received: 30 June 2018; Accepted: 16 August 2018; Published: 18 August 2018

Abstract: Recent developments in humidity sensors have heightened the need for reliability. Seeing as many products such as humidity sensors experience multiple dependent competing failure processes (MDCFPs) with self-recovery, this paper proposes a new general reliability model. Previous research into MDCFPs has primarily focused on the processes of degradation and random shocks, which are appropriate for most products. However, the existing reliability models for MDCFPs cannot fully characterize the failure processes of products such as humidity sensors with significant self-recovery, leading to an underestimation of reliability. In this paper, the effect of self-recovery on degradation was analyzed using a conditional probability. A reliability model for soft failure with self-recovery was obtained. Then, combined with the model of hard failure due to random shocks, a general reliability model with self-recovery was established. Finally, reliability tests of the humidity sensors were presented to verify the proposed reliability model. Reliability modeling for products subject to MDCFPs with considering self-recovery can provide a better understanding of the mechanism of failure and offer an alternative method to predict the reliability of products.

Keywords: reliability model; humidity sensor; self-recovery; dependent competing failure; random shocks

---

1. Introduction

Humidity sensors have been widely used in scientific research and industry applications, such as in the quality control of integrated circuit manufacturing, biological products and pharmaceuticals, and the control of chemical and physical processes [1–7]. The breakdown of humidity sensors may cause the failure of control, detection, and display functions of a system. The rigorous working environment and the diversification of structure and function have put forward increasingly strong reliability requirements for humidity sensors.

The reliability of humidity sensors as an important performance parameter represents the ability of humidity sensors to work without failure under the stated conditions for a specified period. Reliability is a long-term quality indicator for products and cannot be detected before leaving the factory. Reliability modeling is an important tool to evaluate the reliability of products. Reliability modeling for products that experience only soft or hard failure has been extensively explored in the previous studies. Hard failure is when the product's performance remains unchanged before failure and the product suddenly fails at a certain time [8–12], whereas soft failure is the continuous degradation process of a product's performance. When a product's performance exceeds a certain value, soft failure occurs [13–15]. However, due to the complexity of the internal structure and working environment, humidity sensors may deteriorate due to corrosion, fatigue, wear, and other causes. Humidity sensors may also break down suddenly through external shocks. These failure processes compete against each other, and whichever occurs first will cause the humidity sensor to fail. In this case, it is difficult to characterize the failure processes of humidity sensors accurately and comprehensively using soft failure or hard failure alone, which may lead to inaccuracies in the reliability design, analysis, and evaluation. Reliability modeling for humidity sensors and many other products is in line with the actual failure process by combining soft failure and hard failure. Therefore, the reliability theory of competing failure should be used to model the reliability of humidity sensors and many other products.

Competing failure can be categorized into independent and dependent competing failure. In practical applications, competing failure processes are generally dependent on each other. Simply describing the relationship between different failure processes independently often produces an over-estimation of the product's reliability or may even result in unnecessary loss due to untimely maintenance [16–20]. For products subject to multiple dependent competing failure processes (MDCFPs), Peng et al. [21] assumed that some shocks were fatal, which could cause a product's hard failure. Most shocks had little effect on the performance of the product, which could cause sudden damage to continuous performance degradation. In [21], sudden damages were accumulative. Rafiee et al. [22] analyzed a maintenance policy for products subject to MDCFPs and classified random shocks in accordance with the effect of shocks on the failure of products, in which fatal shocks caused hard failure, and non-fatal shocks caused instantaneous damage on degradation. An and Sun [23] discussed a maintenance policy of products subject to MDCFPs and proposed that not all non-fatal shocks caused sudden damages. Only when the amplitudes of shocks were higher than a certain threshold could the shock cause damages. A similar assumption was also found in [24]. Huynh et al. [25] modeled the degradation process through a stochastic process where the degradation process was shown to be strictly increasing. Liu et al. [26] developed a maintenance policy for systems subject to MDCFPs and assumed that the degradation process was an incremental process when system uptime was within a cycle.

Previous literature in this area has some limitations in terms of the reliability research of MDCFPs. Most researchers assumed that sudden damages were accumulative, and that the degradation process strictly increased. This means that previous reliability studies into MDCFPs ignored self-recovery, which is not appropriate for some products. For example, the drift of humidity sensors may undergo a reversible process. External shocks such as rapid humidity increases may cause positive offsets in the long-term continuous drift of humidity sensors. When returning to mild humidity conditions, the offsets slowly decrease. This self-recovery phenomenon exists in many other products and materials, such as mechanics, electronics, micro-electro mechanical system, and self-reconfigurable robotics [27]. After a careful literature review, we found that Liu et al. [28] considered self-recovery and proposed many ideal assumptions regarding self-recovery like the self-recovery process was linear. However, Liu et al. did not develop a specific reliability model with self-recovery. The question of how to characterize the effect of self-recovery reasonably is a challenge that needs to be solved in the reliability modeling for products subject to MDCFPs, and in the reliability analysis of humidity sensors.

In this paper, we developed a new general reliability model for humidity sensors subject to MDCFPs by considering self-recovery. We investigated both hard and soft failure processes. Hard failure is caused by random shocks, whereas soft failure is characterized by a random coefficient regression (RCR) model with positive increments. The RCR model is used to characterize the long-term continuous drift process of humidity sensors, which is caused by physical aging. The positive increments are sudden offsets caused by random shocks. In particular, we took into account that not all non-fatal shocks could cause offsets to the long-term continuous drift. When the inter-arrival time of two continuous shocks is sufficiently large, offsets may decrease. Only when the inter-arrival time of two continuous shocks is under a certain temporal threshold is there a chance that offsets remain. After this, the generality of the developed model is discussed. By setting different parameters, the model can be transformed into different reliability models. Finally, reliability tests of the humidity sensors are given to illustrate the model.

The remainder of this paper is arranged as follows. Section 2 lists the assumptions used in the reliability modeling studies in accordance with the humidity sensors failure process. In Section 3, a reliability model is developed for humidity sensors subject to MDCFPs by considering self-recovery. Moreover, we discuss the generality of the developed model and transform the model into four different reliability models by setting different parameters. In Section 4, reliability tests of the humidity sensors are presented to verify our model. Then, the effects of the parameters on the reliability model are discussed. Section 5 summarizes this paper with concluding remarks.

2. Description of Humidity Sensors Failure Process

The failure of humidity sensors is due to two dependent competing failure processes as shown in Figure 1. The soft failure process is shown in Figure 1a, which is determined by long-term continuous drift, random shocks, and self-recovery. The long-term continuous drift between the humidity sensor's measured values and the actual humidity is caused by physical aging and is often affected by random shocks. When the shock amplitude is less than a certain constant value H, that is, a non-fatal shock, the shock may cause an additional positive offset that arises between the sensor measured value and the actual humidity. The phenomenon wherein shocks can cause increases to degradation exists in many products. In particular, we propose that not all non-fatal shocks cause additional positive offsets. Only frequent shocks with an inter-arrival time of two continuous shocks less than a certain threshold can cause an increase to the long-term continuous drift of humidity sensors. This means that frequent shocks can cause positive offsets, and these offsets are accumulative. If the inter-arrival time of two continuous shocks is larger than a threshold value, once humidity sensors are returned to mild conditions, the offsets may decrease slowly. The self-recovery process is one of decline in positive offsets after a temperature or humidity shock. Self-recovery may be observed once samples are returned to mild environmental conditions. This indicates that this positive drift is a temporary offset that is fully reversible with slow kinetics after returning the sensor to a mild environment. The positive offset is not due to the irreversible damage to the sensing polymer, such as the hydrolysis of the chemical bonds linking the monomers. The positive offset is caused by the rapid environmental change which may self-recover. The hard failure process is shown in Figure 1b. The humidity sensor's exposure to extreme shocks that exceed the threshold level H may cause hard failure. In summary, we assumed that humidity sensors are subject to MDCFPs which include both soft and hard failure. The two competing failure processes are dependent due to the shared exposure to random shocks.

Any of the following conditions cause humidity sensors to fail: (1) the drift of the humidity sensors are beyond the soft failure threshold D, or (2) the magnitude of any shock exceeds the threshold level H.

The specific assumptions used in the reliability modeling in accordance with the humidity sensors failure process can be summarized as follows. The notation used in formulating the reliability models is cited in the Appendix A.

Soft failure occurs when the total drift of the humidity sensors is beyond the failure threshold D. The total drift amount includes the long-term continuous drift with time, positive offsets caused by random shocks, and offset reduction due to self-recovery.

When any shock amplitude exceeds the threshold level H, hard failure occurs.

Shocks occur by a homogeneous Poisson process (HPP), and the rate of HPP is λ. The magnitude of the i-th shock load is denoted as W_i for $i = 1, 2, \ldots, \infty$. W_i is normally distributed $W_i \sim N(\mu_W, \sigma_W)$.

When the inter-arrival time of two continuous shocks is greater than a certain threshold, positive offsets caused by the shocks may decrease, otherwise, positive offsets may remain.

The number of shocks is independent of the magnitude of the shock loads and the positive offsets caused by shocks.

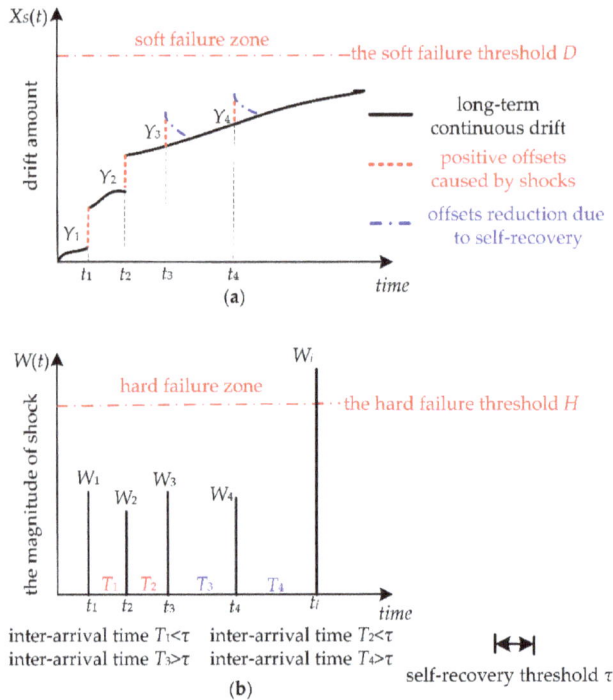

Figure 1. MDCFPs of humidity sensors. (**a**) Soft failure process of humidity sensors; (**b**) Hard failure process of humidity sensors.

3. Reliability Modeling for Humidity Sensors with Considering Self-Recovery

3.1. Reliability Modeling for Humidity Sensors Subject to Soft Failure

Figure 1a demonstrates that the soft failure of humidity sensors occurs when the total drift exceeds D. The total drift $X_S(t)$ includes long-term continuous drift, positive offsets, and offset reduction due to self-recovery. The long-term continuous drift is due to ageing, $X(t)$, is given as

$$X(t) = a + \beta t \tag{1}$$

The $X(t)$ may follow a linear degradation path with random coefficients or a randomized logistic degradation path. Furthermore, it may be necessary to apply a transformation to result in a linear form [29,30]. For illustration purposes, we used a linear degradation path to characterize the long-term continuous drift, where the parameter β is a random variable that corresponds to normal distribution $\beta \sim N(\mu, \sigma^2)$ and where a is a constant.

We assumed that the positive offsets caused by shocks are normally distributed, and denoted as Y_i for $i = 0, 1, 2, \ldots, \infty$, and $Y_i \sim N(\mu_Y, \sigma_Y)$. When considering self-recovery, the positive offsets distribution function is

$$P(\widehat{Y}_i < y) = P(Y_i < y, T_i \le \tau) + P(Y_i < y, T_i > \tau) = (1 - e^{-\lambda \tau})\Phi(y) \tag{2}$$

where $\Phi(\bullet)$ is the cumulative density function (CDF) of a standard normally distributed variable. t_i represents the arrival time of the i-th shock, and T_i represents the inter-arrival time between the i-th shock and the $(i + 1)$-th shock.

The cumulative positive offset $S(t)$ is given by a compound Poisson process

$$S(t) = \begin{cases} \sum_{i=0}^{N(t)} \hat{Y}_i & N(t) > 0 \\ 0 & N(t) = 0 \end{cases} \quad (3)$$

where $N(t)$ is the number of random shocks.

Ignoring self-recovery, the cumulative positive offset $S_1(t)$ can be calculated as

$$S_1(t) = \begin{cases} \sum_{i=0}^{N(t)} Y_i & N(t) > 0 \\ 0 & N(t) = 0 \end{cases} \quad (4)$$

The difference of the cumulative positive offset between considering and ignoring self-recovery can be calculated as

$$S_1(t) - S(t) = \sum_{i=0}^{N(t)} Y_i - (1 - e^{-\lambda \tau}) Y_i = \sum_{i=0}^{N(t)} e^{-\lambda \tau} Y_i N(t) > 0$$

The larger the number of shocks, the higher the shock frequency and the better self-recovery performance, the greater the difference of whether it considers self-recovery.

The probability of the i-th shock occurring by time t is

$$P\{N(t) = i\} = \frac{(\lambda t)^i}{i!} e^{-\lambda t} \quad (5)$$

Furthermore, if we consider $G(t)$ to be the CDF of $\hat{Y}_i$ at time t, and $G^j(t)$ a j convolution of $G(t)$, then the CDF of the $S(t)$ can be derived as

$$P\{S(t) \leq x\} = P\{\sum_{i=0}^{N(t)} \hat{Y}_i \leq x\} = \sum_{j=0}^{N(t)} P\{\sum_{i=0}^{N(t)} \hat{Y}_i \leq x | N(t) = j\} P\{N(t) = j\} = \sum_{j=0}^{N(t)} G^j(x) \frac{(\lambda t)^j e^{-\lambda t}}{j!} \quad (6)$$

The total drift $X_S(t)$ of humidity sensors can be expressed as

$$X_S(t) = X(t) + S(t) \quad (7)$$

By using Equations (1), (3) and (6), the reliability model of humidity sensors subject to soft failure can be derived as

$$F_x(t) = P\{X(t) + S(t) < D\} = P\{X(t) + \sum_{i=0}^{N(t)} \hat{Y}_i < D\} = \sum_{i=0}^{N(t)} P\{(X(t) + \sum_{i=0}^{N(t)} \hat{Y}_i < D) | N(t) = j\} \times P\{N(t) = j\} \quad (8)$$

The reliability model in Equation (8) can be derived for a specific case with a normally distributed Y_i and β

$$R(t) = \Phi(\frac{D - \mu t}{\sigma t}) e^{-\lambda t} + \sum_{i=1}^{N(t)} \Phi(\frac{D - \mu t - i(1 - e^{-\lambda \tau}) \mu_Y}{\sqrt{i(1 - e^{-\lambda \tau})^2 \sigma_Y^2 + \sigma^2 t^2}}) \cdot \frac{(\lambda t)^i e^{-\lambda t}}{i!} = \sum_{i=0}^{N(t)} \Phi(\frac{D - \mu t - i(1 - e^{-\lambda \tau}) \mu_Y}{\sqrt{i(1 - e^{-\lambda \tau})^2 \sigma_Y^2 + \sigma^2 t^2}}) \cdot \frac{(\lambda t)^i e^{-\lambda t}}{i!} \quad (9)$$

3.2. Reliability Modeling for Humidity Sensors Subject to Random Shocks

When the shock load exceeds the threshold level H, hard failure occurs. According to the stress-strength model [31], the probability of surviving the i-th shock is shown as

$$P(W_i < H) = F_W(H) \ i = 1, 2, \dots, N(t). \quad (10)$$

In this paper, a stochastic extreme shock model was used to characterize the random shocks that cause hard failure. The magnitude of the i-th shock is denoted as W_i for $i = 1, 2, \ldots, N(t)$, and $W_i \sim N(\mu_W, \sigma_W)$. Therefore, the probability of survival in Equation (10) is

$$F_W(H) = P(W_i < H) = \Phi\left(\frac{H - \mu_W}{\sigma_W}\right) i = 1, 2, \ldots, N(t). \tag{11}$$

where the $\Phi(\bullet)$ is the CDF of a standard normally distributed variable.

3.3. Reliability Modeling for Humidity Sensors Subject to MDCFPs

Hard or soft failure can cause humidity sensors to fail. The reliability function can be derived as

$$
\begin{aligned}
R(t) &= P(X(t) < D, N(t) = 0) + \sum_{i=1}^{N(t)} P(W_1 < H, \ldots, W_{N(t)} < H, X(t) + \sum_{i=1}^{N(t)} \hat{Y}_i < D, N(t) = i) \\
&= P(X(t) < D, N(t) = 0) + \sum_{i=1}^{N(t)} F_W(H)^i P(X(t) + \sum_{i=1}^{N(t)} \hat{Y}_i < D)|N(t) = i) \times P\{N(t) = i\}
\end{aligned} \tag{12}
$$

The reliability function can be expressed for a more specific case

$$
\begin{aligned}
R(t) &= P(X(t) < D, N(t) = 0) + \sum_{i=1}^{N(t)} P(W_1 < H, \ldots, W_{N(t)} < H, X(t) + \sum_{i=1}^{N(t)} \hat{Y}_i < D, N(t) = i) \\
&= \Phi\left(\frac{D - \mu t}{\sigma t}\right) e^{-\lambda t} + \sum_{i=1}^{N(t)} \Phi\left(\frac{D - \mu t - i(1 - e^{-\lambda \tau})\mu_Y}{\sqrt{\sigma^2 t^2 + i(1 - e^{-\lambda \tau})^2 \sigma_Y^2}}\right) \cdot \frac{(\lambda t)^i e^{-\lambda t}}{i!} \cdot \left[\Phi\left(\frac{H - \mu_W}{\sigma_W}\right)\right]^i \\
&= \sum_{i=0}^{N(t)} \Phi\left(\frac{D - \mu t - i(1 - e^{-\lambda \tau})\mu_Y}{\sqrt{\sigma^2 t^2 + i(1 - e^{-\lambda \tau})^2 \sigma_Y^2}}\right) \cdot \frac{(\lambda t)^i e^{-\lambda t}}{i!} \cdot \left[\Phi\left(\frac{H - \mu_W}{\sigma_W}\right)\right]^i
\end{aligned} \tag{13}
$$

As shown in Equation (13), when $0 < \tau < \infty$, the smaller the value of τ, the stronger the product's self-recovery. When $\tau_1 < \tau_2$, offsets caused by shocks with an inter-arrival time greater than τ_1 can recover, which includes the offsets caused by shocks with the inter-arrival time between τ_1 and τ_2. Therefore, the smaller the value of τ, the more the offsets recover, the less the degradation volume, and the higher the reliability.

Based on Equation (13), the probability density function (PDF) of the failure time is

$$
\begin{aligned}
f(t) &= -\frac{dR(t)}{dt} = -\sum_{i=1}^{N(t)} \left[\Phi\left(\frac{H - \mu_W}{\sigma_W}\right)\right]^i \phi\left(\frac{D - \mu t - i(1 - e^{-\lambda \tau})\mu_Y}{\sqrt{\sigma^2 t^2 + i(1 - e^{-\lambda \tau})\sigma_Y^2}}\right) \\
&\times \left(\frac{-\mu(\sigma^2 t^2 + (1 - e^{-\lambda \tau})\sigma_Y^2) - \sigma^2 t(D - \mu t - i(1 - e^{-\lambda \tau})\mu_Y)}{(\sigma^2 t^2 + i(1 - e^{-\lambda \tau})\sigma_Y^2)^{\frac{3}{2}}}\right) \times \frac{(\lambda t)^i e^{-\lambda t}}{i!} \\
&- \sum_{i=1}^{N(t)} \left[\Phi\left(\frac{H - \mu_W}{\sigma_W}\right)\right]^i \Phi\left(\frac{D - \mu t - i(1 - e^{-\lambda \tau})\mu_Y}{\sqrt{\sigma^2 t^2 + i(1 - e^{-\lambda \tau})\sigma_Y^2}}\right) \times \frac{\lambda(\lambda t)^{i-1} e^{-\lambda t}(-\lambda t + i)}{i!} \\
&- \phi\left(\frac{D - \mu t}{\sigma t}\right) \times \left(-\frac{D}{\sigma t^2}\right) e^{-\lambda t} + \lambda \Phi\left(\frac{D - \mu t}{\sigma t}\right) e^{-\lambda t}
\end{aligned} \tag{14}
$$

where $\phi(\bullet)$ is the PDF of a standard normally distributed variable.

3.4. Some Special Cases

With different parameters, the reliability model Equation (13) can be transformed into different reliability models and coincides with models with a slight difference to the previous literature.

When $\tau = \infty$, the reliability model Equation (13) can be transformed into a reliability model for dependent competing failure as shown in Equation (15). The model of Equation (15) ignores self-recovery as with the previous literature [21]. As $\tau = \infty$ means that when the inter-arrival time of two continuous shocks is smaller than infinite, a shock can cause positive offsets to the continuous long-term drift, that is, all shocks can cause offsets.

Ignoring self-recovery ($\tau = \infty$), the reliability is shown as

$$
\begin{aligned}
R(t) &= P(X(t) < D, N(t) = 0) + \sum_{i=1}^{N(t)} P(W_1 < H, \ldots, W_{N(t)} < H, X(t) + \sum_{i=1}^{N(t)} Y_i < D, N(t) = i) \\
&= \Phi(\tfrac{D-\mu t}{\sigma t}) e^{-\lambda t} + \sum_{i=1}^{N(t)} \Phi(\tfrac{D-\mu t - i\mu_Y}{\sqrt{\sigma^2 t^2 + i\sigma_Y^2}}) \cdot \tfrac{(\lambda t)^i e^{-\lambda t}}{i!} \cdot [\Phi(\tfrac{H-\mu_W}{\sigma_W})]^i \\
&= \sum_{i=0}^{N(t)} \Phi(\tfrac{D-\mu t - i\mu_Y}{\sqrt{\sigma^2 t^2 + i\sigma_Y^2}}) \cdot \tfrac{(\lambda t)^i e^{-\lambda t}}{i!} \cdot [\Phi(\tfrac{H-\mu_W}{\sigma_W})]^i
\end{aligned}
\tag{15}
$$

Based on Equation (15), the PDF of the failure time is derived as

$$
\begin{aligned}
f(t) &= -\frac{dR(t)}{dt} = -\sum_{i=1}^{N(t)} [\Phi(\tfrac{H-\mu_W}{\sigma_W})]^i \phi(\tfrac{D-\mu t - i\mu_Y}{\sqrt{\sigma^2 t^2 + i\sigma_Y^2}}) \times (\tfrac{-\mu(\sigma^2 t^2 + i\sigma_Y^2) - \sigma^2 t(D-\mu t - i\mu_Y)}{(\sigma^2 t^2 + i\sigma_Y^2)^{\frac{3}{2}}}) \\
&\times \tfrac{(\lambda t)^i e^{-\lambda t}}{i!} - \sum_{i=1}^{N(t)} [\Phi(\tfrac{H-\mu_W}{\sigma_W})]^i \phi(\tfrac{D-\mu t - i\mu_Y}{\sqrt{\sigma^2 t^2 + i\sigma_Y^2}}) \times \tfrac{\lambda(\lambda t)^{i-1} e^{-\lambda t}(-\lambda t + i)}{i!} - \phi(\tfrac{D-\mu t}{\sigma t}) \times (-\tfrac{D}{\sigma t^2}) e^{-\lambda t} \\
&+ \lambda \Phi(\tfrac{D-\mu t}{\sigma t}) e^{-\lambda t}
\end{aligned}
\tag{16}
$$

When $\tau = 0$, the reliability model of Equation (13) is transformed into a reliability model for independent competing failure as shown in Equation (17). As $\tau = 0$ means that positive offsets can recover when the inter-arrival time of two continuous shocks is greater than 0, that means all offsets can recover. It also means that shocks do not cause offsets to long-term continuous drift when $\tau = 0$, that is, hard failure and soft failure are independent of each other. This reliability model is similar to the model used in the previous study [32].

When soft failure and hard failure are independent ($\tau = 0$), the reliability is shown as

$$
\begin{aligned}
R(t) &= P(X(t) < D, N(t) = 0) + \sum_{i=1}^{N(t)} P(W_1 < H, \ldots, W_{N(t)} < H, X(t) < D, N(t) = i) \\
&= \Phi(\tfrac{D-\mu t}{\sigma t}) e^{-\lambda t} + \sum_{i=1}^{N(t)} \Phi(\tfrac{D-\mu t}{\sigma t}) \cdot \tfrac{(\lambda t)^i e^{-\lambda t}}{i!} \cdot [\Phi(\tfrac{H-\mu_W}{\sigma_W})]^i = \sum_{i=0}^{N(t)} \Phi(\tfrac{D-\mu t}{\sigma t}) \cdot \tfrac{(\lambda t)^i e^{-\lambda t}}{i!} \cdot [\Phi(\tfrac{H-\mu_W}{\sigma_W})]^i
\end{aligned}
\tag{17}
$$

Based on Equation (17), the PDF of the failure time is derived as

$$
\begin{aligned}
f(t) &= -\frac{dR(t)}{dt} = -\sum_{i=0}^{N(t)} [\Phi(\tfrac{H-\mu_W}{\sigma_W})]^i \times \phi(\tfrac{D-\mu t}{\sigma t}) \times (-\tfrac{D}{\sigma t^2}) \tfrac{(\lambda t)^i e^{-\lambda t}}{i!} \\
&- \sum_{i=0}^{N(t)} [\Phi(\tfrac{H-\mu_W}{\sigma_W})]^i \times \Phi(\tfrac{D-\mu t}{\sigma t}) \times \tfrac{\lambda(\lambda t)^{i-1} e^{-\lambda t}(-\lambda t + i)}{i!}
\end{aligned}
\tag{18}
$$

Reliability modeling for products that experience soft failure only concerns the performance degradation process. By setting the parameters of random shock in Equation (13) to 0 and ignoring hard failure, the reliability model of Equation (13) can be converted to the reliability model based on performance degradation.

$$
R(t) = P(X(t) < D) = \Phi(\frac{D-\mu t}{\sigma t})
\tag{19}
$$

Based on Equation (19), the PDF of the failure time is derived as

$$
f(t) = \phi(\frac{D-\mu t}{\sigma t})(-\frac{D}{\sigma t^2})
\tag{20}
$$

Traditional reliability theory only focuses on hard failure. By setting the parameters of soft failure in Equation (13) to 0, the reliability model of Equation (13) can be converted to the traditional reliability model, which only considers hard failure due to random shocks.

Sensors **2018**, *18*, 2714

$$R(t) = \sum_{i=0}^{N(t)} [\Phi(\frac{H - \mu_W}{\sigma_W})]^i \frac{(\lambda t)^i e^{-\lambda t}}{i!} \tag{21}$$

Based on Equation (21), the PDF of the failure time is derived as

$$f(t) = \sum_{i=1}^{N(t)} [\Phi(\frac{H - \mu_W}{\sigma_W})]^i \times \frac{\lambda(\lambda t)^{i-1} e^{-\lambda t}(-\lambda t + i)}{i!} - \lambda e^{-\lambda t} \tag{22}$$

The reliability model Equation (13) developed in this paper can be transformed into different reliability models seen in previous literature, as shown in Table 1. This means that models 1, 2, 3, and 4 are special cases of the reliability model developed in this paper.

Table 1. Reliability models.

Model	Description of the Failure Process	Expression
Model 1	This model characterizes hard failure which is caused by a stochastic shock process.	Equation (21)
Model 2	This model characterizes soft failure process. Products may not be subject to random shocks.	Equation (19)
Model 3	This model characterizes independent competing failure processes. Soft failure and hard failure are independent.	Equation (17)
Model 4	This model characterizes MDCFPs but ignores self-recovery. All non-fatal shocks can cause sudden increases in degradation.	Equation (15)
Model proposed in this paper	This model characterizes MDCFPs by considering self-recovery and can be transformed into four different reliability models (Models 1, 2, 3, and 4) by varying parameters.	Equation (13)

4. Numerical Examples and Results

The following two examples in this section are presented to illustrate the model discussed in the previous section.

4.1. Example I

A humidity sensors reliability test conducted at AMS Netherlands BV Laboratories was used here to verify the proposed model [33]. Ageing of the sensor may cause a measured value long-term drift. This long-term drift is a continuous degradation process and is often affected by random shocks. When subject to random shocks, such as a rapid increase in humidity, positive offsets may be caused to the long-term continuous drift, especially when humidity sensors return to mild humidity conditions for a long time, that is, if the inter-arrival time of two continuous shocks is long enough, positive offsets will slowly decrease. In contrast, if the inter-arrival times are less than a certain value, positive offsets may remain.

To illustrate the model developed in this paper, we set the parameters shown in Table 2. The reliability model in this paper was based on the statistical analysis of a pseudo failure life and model of hard failure due to random shocks. The pseudo failure life was obtained by extrapolating the degradation path. In the linear degradation path $X(t) = a + \beta t$, the distribution of β can be obtained by recording the drift data of the humidity sensor. It was assumed that β is a normally distributed random variable, that is, the degradation volume at any time t following a normal distribution. We assumed that the humidity sensors did not degenerate at the initial time, that is, $a = 0$. From the test results, we also obtained the soft failure threshold D and the hard failure threshold H. We assumed the size of the shock loads as W_i following a distribution, and consequently the positive offset Y_i also followed a normal distribution.

Table 2. Parameters of the reliability model for example I.

Parameter	Value	Description
D	10	The soft failure threshold is 10: when the drift amount rises by 10%, soft failure occurs.
H	85	The hard failure threshold is 85: when relative humidity exceeds 85%, hard failure occurs.
β	$N(0.0005, 0.005^2)$	The drift rate.
α	$\alpha = 0$	The drift value at the initial time ($t = 0$).
W_i	$N(65, 8^2)$	The i-th shock amplitude.
λ	0.02/h	The rate of a homogeneous Poisson random shock process.
Y_i	$N(0.2, 0.02^2)$	The positive offsets caused by the i-th shock.

The four different reliability models used in previous studies are special cases of the proposed model as discussed in Section 3.4. All models are drawn and compared in Figure 2. We found that when considering only hard or soft failure (Models 1 and 2), the reliabilities were higher than when both failures had a competitive relationship (Models 3, 4, and the proposed model). For three reliability models of competing failure, Models 3 and 4 were special cases of the proposed model. The reliabilities of the competing failure processes had a similar trend of change. Between 1000 h and 5000 h, the reliability of the independent competing failure (Model 3) was higher than the reliability of the dependent competing failure with self-recovery (the proposed model). When it was assumed that the relationship between soft failure and hard failure was independent, random shock did not affect the long-term continuous drift of the humidity sensors. The degradation amount of the independent competing failure was less than the degradation amount of the dependent competing failure with self-recovery. If we assumed that the failure processes were independent, then the computed reliability might be higher than the actual reliability of the humidity sensors. At the same time, the degradation amount of the dependent competing failure with self-recovery (the proposed model) was less than the degradation amount of the dependent competing failure without considering self-recovery (Model 4). If we ignored the self-recovery processes, then the computed reliability might be smaller than the actual reliability of the humidity sensors. Therefore, for the reliability modeling of some products such as humidity sensors, the self-recovery processes need to be considered.

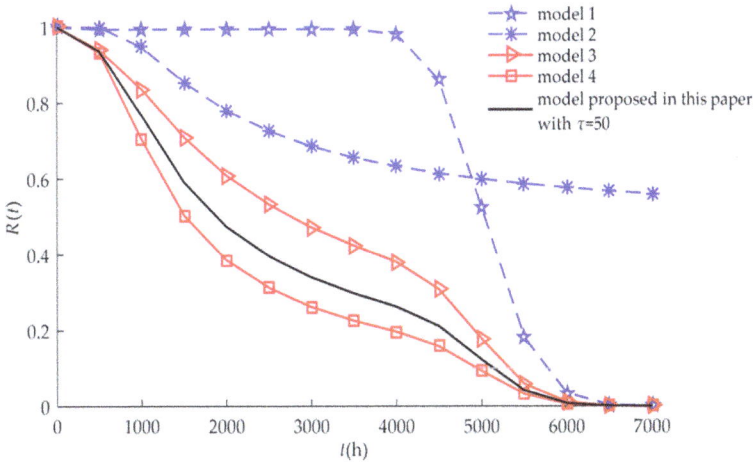

Figure 2. Comparison of $R(t)$ for different models for example I.

The failure rate functions of Equations (14), (16), (18), (20), and (22) are shown in Figure 3. When only hard failure was considered (Model 1), the failure rate increased significantly after 4000 h. In contrast, when only soft failure was considered (Model 2), the rate was mainly concentrated prior to 4000 h. For the three competing models (Models 3, 4, and the proposed model), the failure rate functions were almost non-zero throughout the service life of the humidity sensors, that is, the humidity sensors could fail at any time during service. When only soft failure or hard failure was considered, the humidity sensors could fail within a specified time.

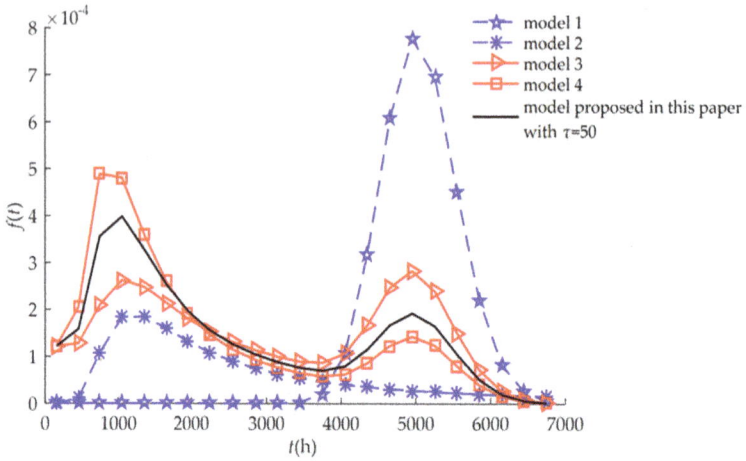

Figure 3. Comparison of $f(t)$ for different models for example I.

To explore the influence of parameters on the reliability of humidity sensors, sensitivity analyses of $R(t)$ on τ, D, H, λ are presented in Figures 4–7 respectively for example I.

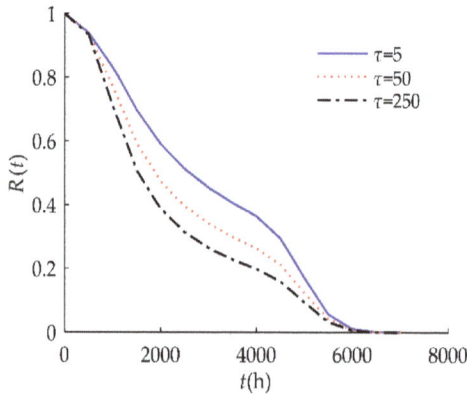

Figure 4. Sensitivity analysis of $R(t)$ on τ for example I.

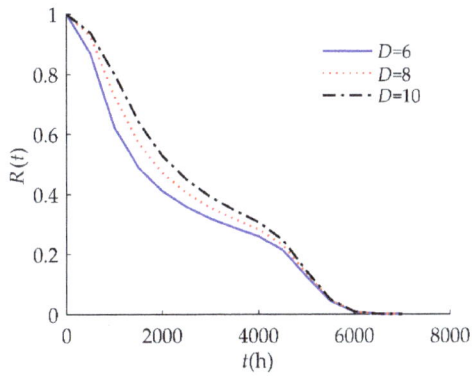

Figure 5. Sensitivity analysis of $R(t)$ on D for example I.

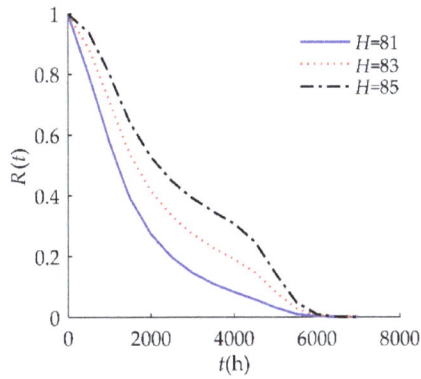

Figure 6. Sensitivity analysis of $R(t)$ on H for example I.

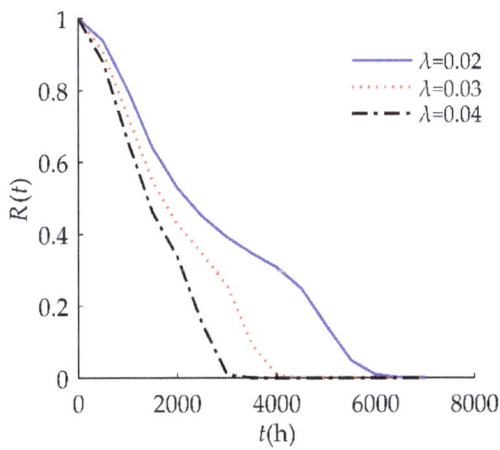

Figure 7. Sensitivity analysis of $R(t)$ on λ for example I.

Figure 4 indicates that the self-recovery threshold τ had a significant effect on $R(t)$. When τ decreased, the reliability of the humidity sensor increased after 1000 h. As discussed in Section 3.3, the smaller the value of τ, the higher the reliability of the products.

Figure 5 indicates that $R(t)$ as sensitive to the soft failure threshold D. When D decreased from 10 to 6, $R(t)$ decreased, which means that the reliability is lower when D gets smaller.

In Figure 6, the hard failure threshold H had an obvious effect on $R(t)$. When the hard failure threshold H decreased, $R(t)$ decreased. An explanation for this may be that products with a higher H have a better ability to resist shock.

In Figure 7, we observed that $R(t)$ was susceptible to the random shock rate λ. When λ decreased, $R(t)$ shifted to the right. This result indicates that a larger λ decreases reliability performance. An explanation for this might be that the higher the shock rate, the more positive offsets on the degradation value, so failure occurs at a much earlier time.

4.2. Example II

A case study of a solid state relative humidity (RH) sensor reliability analysis by the University of Wisconsin-Madison is provided to illustrate the model [34]. The drift of capacitance–RH characteristic is the dominant failure mode for a solid state RH sensor at 85 °C/85% RH. In the temperature test of a solid state RH sensor, the sensor breaks down when the magnitude of the temperature shock is above a certain level. In addition, positive offsets to the drift of the capacitance–RH characteristic are caused when the temperature shock is non-fatal. In particular, when the humidity sensors return to mild conditions for a long time, the positive offsets slowly decrease. To illustrate the model developed in this paper, we set the parameters shown in Table 3. The linear degradation path was $X(t) = a + \beta t$, where $a = 0$ and β is normally distributed was obtained by the test data. The shock size and positive offsets caused by the shocks were assumed to be normally distributed.

Table 3. Parameters of the reliability model for example II.

Parameter	Value	Description
D	10	The soft failure threshold is 10: when the drift amount rises by 10%, soft failure occurs.
H	95	The hard failure threshold is 95: when ambient temperature exceeds 95 °C, hard failure occurs.
β	$N(0.0893, (0.0090)^2)$	The drift rate of capacitance–RH characteristic at 85 °C/85% RH.
α	$a = 0$	The drift value at the initial time ($t = 0$).
W_i	$N(85, 8^2)$	The i-th shock amplitude.
λ	0.1/Day	The rate of a homogeneous Poisson random shock process.
Y_i	$N(0.2, 0.02^2)$	The positive offsets caused by the i-th shock.

The four different reliability models (Models 1, 2, 3, and 4) used in previous studies are special cases of the proposed model as discussed in Section 3.4. All models are drawn and compared in Figure 8. The corresponding failure rate functions are shown in Figure 9. We also found that when considering only hard or soft failure, the reliabilities were higher than that when both failures had a competitive relationship, the failure rate functions were not 0 within a certain time, and not all the whole time of service. The three reliabilities of competing failure had some of the same change trends. The reliability of the independent competing is the highest, followed by the reliability of the dependent competing with considering self-recovery.

To explore the influence of the parameters on the reliability of humidity sensors, the sensitivity analyses of $R(t)$ on τ, D, H, and λ are presented in Figures 10–13, respectively for example II. These indicate that the reliability performance was better for a smaller self-recovery threshold τ, larger soft failure threshold D, larger hard failure threshold H, or smaller random shock rate λ.

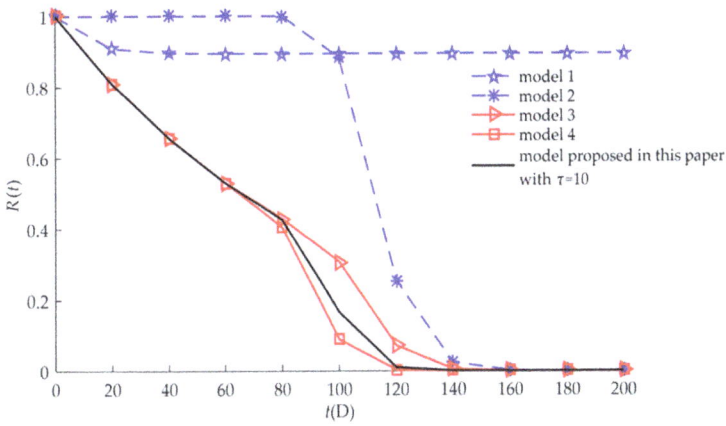

Figure 8. Comparison of $R(t)$ for different models for example II.

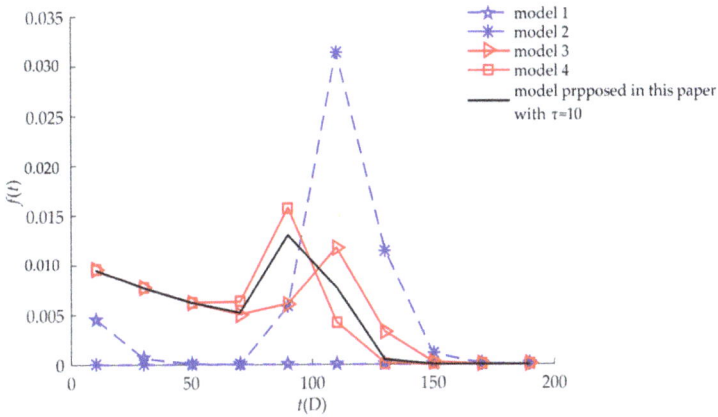

Figure 9. Comparison of $f(t)$ for different models for example II.

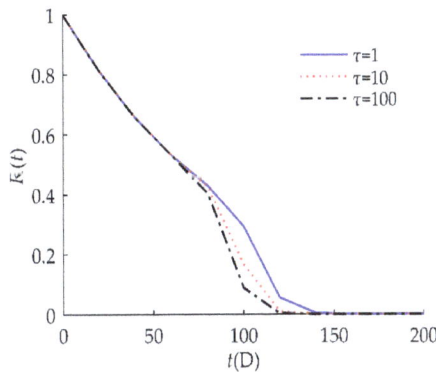

Figure 10. Sensitivity analysis of $R(t)$ on τ for example II.

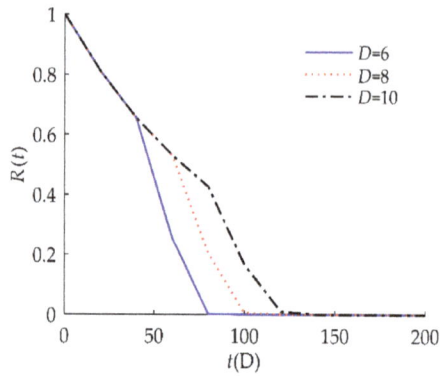

Figure 11. Sensitivity analysis of $R(t)$ on D for example II.

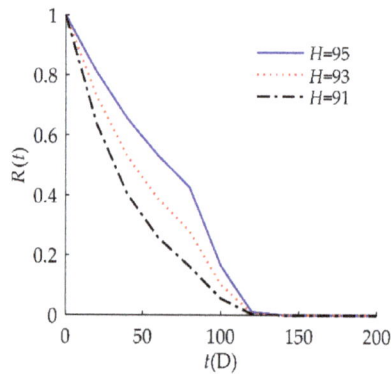

Figure 12. Sensitivity analysis of $R(t)$ on H for example II.

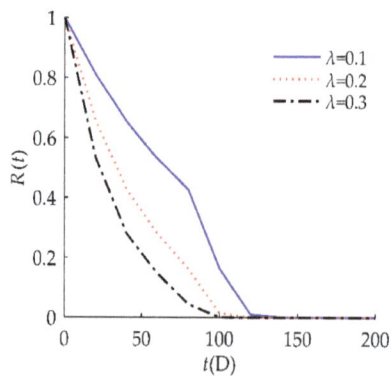

Figure 13. Sensitivity analysis of $R(t)$ on λ for example II.

5. Conclusions

In this paper, we proposed a new and more general reliability model for humidity sensors subjected to dependent competing failure with considering self-recovery. This paper analyzed the condition of self-recovery, that is, it focused on the effect of the inter-arrival time of shocks on continuous degradation. On this basis, a reliability model for soft failure that considered self-recovery was established. Combined with the reliability analysis of hard failure due to shocks, a new reliability model for dependent competing failure that considered self-recovery was developed. By adjusting the different parameters, the generality of the developed model was discussed. It was found that the four different reliability models used in previous studies were the special cases of the model developed in this paper. This new model represents a major extension on previous studies. We presented examples to demonstrate the reliability model and analyzed the effects of the parameters on reliability. For further studies, additional terms of the self-recovery condition can be considered, such as the magnitude of the shocks.

Author Contributions: Z.Z. supervised the research. J.Q. analyzed the data and wrote the paper. C.N. contributed to the literature review and helped to perform reliability analysis. C.W. analyzed the experiments and compiled the program. J.W. reviewed and edited the manuscript. All authors read and approved the final manuscript.

Funding: This research received no external funding.

Acknowledgments: This study was supported by Heilongjiang Provincial Fund Grant (Grant No. QC2016068) and the National Natural Science Foundation of China Youth Fund Project (Grant No. 61501149).

Conflicts of Interest: The authors declare no conflict of interest.

Appendix A

The notation used in formulating the reliability models is now listed.

t	time
$X(t)$	drift of the measured value compared to reference caused by physical ageing at time t
$N(t)$	the number of shocks
Y_i	a positive offset between the sensor measured value and the actual conditions caused by the i-th shock
W_i	the magnitude of the i-th shock
D	the threshold of soft failure
H	the threshold of hard failure
λ	the rate of a homogeneous Poisson random shock process
$X_S(t)$	total drift volume at time t composed to long-term continuous drift, positive offsets caused by random shocks, and offset reduction due to self-recovery
$R(t)$	time-dependent reliability
T_i	inter-arrival time between the i-th shock and the $(i + 1)$-th shock
τ	the self-recovery threshold
$S(t)$	cumulative positive offset caused by random shocks at time t

References

1. Hernandezrivera, D.; Rodriguezroldan, G.; Moramartinez, R.; Suastegomez, E. A Capacitive Humidity Sensor Based on an Electrospun PVDF/Graphene Membrane. *Sensors* **2017**, *17*, 1009. [CrossRef] [PubMed]
2. Dessler, A.E.; Sherwood, S.C. A Matter of Humidity. *Science* **2009**, *323*, 1020–1021. [CrossRef] [PubMed]
3. Liehr, S.; Breithaupt, M.; Krebber, K. Distributed Humidity Sensing in PMMA Optical Fibers at 500 nm and 650 nm Wavelengths. *Sensors* **2017**, *17*, 738. [CrossRef] [PubMed]
4. Previati, M.; Canone, D.; Bevilacqua, I.; Boetto, G.; Pognant, D.; Ferraris, S. Evaluation of wood degradation for timber check dams using time domain reflectometry water content measurements. *Ecol. Eng.* **2012**, *44*, 259–268. [CrossRef]
5. Boudaden, J.; Steinmabl, M.; Endres, H.E.; Drost, A.; Eisele, I.; Kutter, C.; Muller-Buschbaum, P. Polyimide-Based Capacitive Humidity Sensor. *Sensors* **2018**, *18*, 1516. [CrossRef] [PubMed]
6. Park, H.; Lee, S.; Jeong, S.H.; Jung, U.H.; Park, H.; Lee, M.G.; Kim, S.; Lee, J. Enhanced Moisture-Reactive Hydrophilic-PTFE-Based Flexible Humidity Sensor for Real-Time Monitoring. *Sensors* **2018**, *18*, 921. [CrossRef] [PubMed]
7. Blank, T.A.; Eksperiandova, L.P.; Belikov, K.N. Recent trends of ceramic humidity sensors development: A review. *Sens. Actuators B Chem.* **2016**, *228*, 416–442. [CrossRef]

8. Zuo, M.J.; Jiang, R.; Yam, R.C.M. Approaches for reliability modeling of continuous-state devices. *IEEE Trans. Reliab.* **2002**, *48*, 9–18. [CrossRef]

9. Barnett, T.S.; Grady, M.; Purdy, K.; Singh, A.D. Exploiting defect clustering for yield and reliability prediction. *IEE Proc.-Comput. Dig. Tech.* **2005**, *152*, 407–414. [CrossRef]

10. Huang, W.; Askin, R.G. A generalized SSI reliability model considering stochastic loading and strength aging degradation. *IEEE Trans. Reliab.* **2004**, *53*, 77–82. [CrossRef]

11. Mallor, F.; Santos, J. *Classification of Shock Models in System Reliability*; VII Jornadas Zaragoza-Pau de Matemática Aplicada y Estadística: Jaca (Huesca), 17–18 de Septiembre de 2001; Prensas Universitarias de Zaragoza: Zaragoza, Spain, 2003; pp. 405–412.

12. Fan, J.; Ghurye, S.G.; Levine, R.A. Multicomponent Lifetime Distributions in the Presence of Ageing. *J. Appl. Probab.* **2000**, *37*, 521–533. [CrossRef]

13. Lu, C.J.; Meeker, W.Q. Using Degradation Measures to Estimate a Time-to-Failure Distribution. *Technometrics* **1993**, *35*, 161–174. [CrossRef]

14. Kharoufer, J.P.; Cox, S.M. Stochastic models for degradation-based reliability. *IIE Trans.* **2005**, *37*, 533–542. [CrossRef]

15. Park, C.; Padgett, W.J. Stochastic degradation models with several accelerating variables. *IEEE Trans. Reliab.* **2006**, *55*, 379–390. [CrossRef]

16. Chien, Y.H.; Shen, S.H.; Zhang, Z.G.; Love, E. An extended optimal replacement model of systems subject to shocks. *Eur. J. Oper. Res.* **2006**, *175*, 399–412. [CrossRef]

17. Keedy, E.; Feng, Q. A physics-of-failure-based reliability and maintenance modeling framework for stent deployment and operation. *Reliab. Eng. Syst. Saf.* **2012**, *103*, 94–101. [CrossRef]

18. Li, W.; Pham, H. An inspection-maintenance model for systems with multiple competing processes. *IEEE Trans. Reliab.* **2005**, *54*, 318–327. [CrossRef]

19. Cha, J.H.; Pulcini, G. A Dependent Competing Risks Model for Technological Units Subject to Degradation Phenomena and Catastrophic Failures. *Qual. Reliab. Eng. Int.* **2016**, *32*, 505–517. [CrossRef]

20. Wang, Y.; Pham, H. A Multi-Objective Optimization of Imperfect Preventive Maintenance Policy for Dependent Competing Risk Systems with Hidden Failure. *IEEE Trans. Reliab.* **2011**, *60*, 770–781. [CrossRef]

21. Peng, H.; Feng, Q.; Coit, D.W. Reliability and maintenance modeling for systems subject to multiple dependent competing failure processes. *IIE Trans.* **2010**, *43*, 12–22. [CrossRef]

22. Rafiee, K.; Feng, Q.; Coit, D.W. Condition-based maintenance for repairable deteriorating systems subject to generalized mixed shock model. *IEEE Trans. Reliab.* **2015**, *64*, 1164–1174. [CrossRef]

23. An, Z.; Sun, D. Reliability modeling for systems subject to multiple dependent competing failure processes with shock loads above a certain level. *Reliab. Eng. Syst. Saf.* **2017**, *157*, 129–138. [CrossRef]

24. Rafiee, K.; Feng, Q.; Coit, D.W. Reliability assessment of competing risks with generalized mixed shock models. *Reliab. Eng. Syst. Saf.* **2017**, *159*, 1–11. [CrossRef]

25. Huynh, K.T.; Barros, A.; Berenguer, C.; Castro, I.T. A periodic inspection and replacement policy for systems subject to competing failure modes due to degradation and traumatic events. *Reliab. Eng. Syst. Saf.* **2011**, *96*, 497–508. [CrossRef]

26. Liu, X.; Li, J.; Alkhalifa, K.N.; Hamouda, A.S.; Coit, D.W.; Elsayed, E.A. Condition-based maintenance for continuously monitored degrading systems with multiple failure modes. *IIE Trans.* **2013**, *45*, 422–435. [CrossRef]

27. Regina, F.; Richard, M.C.; Benjamin, D.; Alan, P.; Asutosh, T.; Giovanna, D.M.S. Self-healing and self-repairing technologies. *Int. J. Adv. Manuf. Technol.* **2013**, *69*, 1033–1061.

28. Liu, H.; Yeh, R.H.; Cai, B. Reliability modeling for dependent competing failure processes of damage self-healing systems. *Comput. Ind. Eng.* **2017**, *105*, 55–62. [CrossRef]

29. Wang, Y.; Pham, H. Imperfect preventive maintenance policies for two-process cumulative damage model of degradation and random shocks. *Int. J. Syst. Assur. Eng. Manag.* **2011**, *2*, 66–77. [CrossRef]

30. Bae, S.J.; Kou, W.; Kvam, P.H. Degradation models and implied lifetime distribution. *Reliab. Eng. Syst. Saf.* **2007**, *92*, 601–608. [CrossRef]

31. Rafiee, K.; Feng, Q.; Coit, D.W. Reliability modeling for dependent competing failure processes with changing degradation rate. *IIE Trans.* **2014**, *46*, 483–496. [CrossRef]

32. Wang, Z.; Huang, H.; Li, Y.; Xiao, N. An approach to reliability assessment under degradation and shock process. *IEEE Trans. Reliab.* **2011**, *60*, 852–863. [CrossRef]

Sensors **2018**, *18*, 2714

33. Jose, S.; Vooge, F.; Schaar, C.V.D.; Nath, S.; Nenadovic, N.; Vanhelmont, F.; Lous, E.J.; Suy, H.; Zandt, M.I.; Sakic, A.; et al. Reliability tests for modeling of relative humidity sensor drifts. In Proceedings of the 2017 IEEE Reliability Physics Symposium, Monterey, CA, USA, 2–6 April 2017.

34. Denton, D.D.; Jaafar, M.A.S.; Ralston, A.R.K. The long term reliability of a switched-capacitor relative humidity sensor system. In Proceedings of the IEEE International Symposium on Circuits and Systems, San Diego, CA, USA, 10–13 May 1992; Volume 4, pp. 1840–1843.

MDPI

St. Alban-Anlage 66

4052 Basel

Switzerland

Tel. +41 61 683 77 34

Fax +41 61 302 89 18

www.mdpi.com

Sensors Editorial Office

E-mail: sensors@mdpi.com

www.mdpi.com/journal/sensors

www.ingramcontent.com/pod-product-compliance
Lightning Source LLC
Chambersburg PA
CBHW051852210326
41597CB00033B/5869